THE
VITAMIN
COMPLEX

THE
VITAMIN
COMPLEX

OUR OBSESSIVE QUEST FOR
NUTRITIONAL PERFECTION

CATHERINE PRICE

ONEWORLD

A Oneworld book

First published in Great Britain and the Commonwealth by Oneworld
Publications, 2015

First published in the US by Penguin Press, a member of the Penguin
Group (USA) Inc.

ISBN 978-1-78074-346-2
eISBN 978-1-78074-347-9

Printed and bound in Denmark by Nørhaven

Oneworld Publications
10 Bloomsbury Street
London WC1B 3SR
England

Stay up to date with the latest books,
special offers, and exclusive content from
Oneworld with our monthly newsletter

Sign up on our website
www.oneworld-publications.com

MIX
Paper from
responsible sources
FSC® C104608

For Peter and my parents

The more we know, the more we don't know.

—A NUTRITIONIST AT THE US DEPARTMENT OF AGRICULTURE, 2013

CONTENTS

INTRODUCTION

On the morning of June 12, 1769, a young British physician named William Stark began a rather peculiar experiment. Six feet tall, with red hair and a rosy complexion, Stark lived in London during a time when repetitive menus of bland foods were thought to be the healthiest way to eat. Stark, like many of us, preferred a more "pleasant and varied diet," and wanted to know whether it was really necessary to be so boring. So he designed a series of studies to test this theory, and recruited an extraordinarily eager and willing test subject: himself.

Stark, who appears to have been rather impressionable, had gathered many stories from his mentors about the benefits of eating simply. Sir John Pringle, a Scottish physician often referred to as the father of modern military medicine, told Stark of a Greek island whose inhabitants "lived wholly on currants," and of a 90-year-old woman who ate "only the pure fat of meat." Benjamin Franklin boasted to Stark that during his days as a printer, he had once survived for two weeks on bread and water and was "stout and hearty with this diet." Stark hoped he might prove his mentors wrong, but he promised to remain objective. "I confess it will afford me a singular pleasure if I can prove, by experiment, that a pleasant and varied diet is equally conducive to health," he admitted. "[But] if a simpler diet seems the most healthy, I shall not hesitate to declare it."

After conducting some preliminary trials with mustard and caraway seeds to see how long food took to pass through his body, Stark began his formal experimentation with a Franklin-inspired meal plan: bread, water, and, after a time, sugar. Nothing else. In his journal, he kept fastidious records of every variable he thought might be relevant to his health, weighing his food and excretions, tracking the weather, and even distinguishing time spent "buttoned up in [his] great coat" from time passed sitting "quite naked by the fire."

Whereas Franklin only lasted two weeks, Stark stuck to this diet—which was both incredibly restrictive and completely lacking in fresh produce—for almost two months. By August 11, he reported small ulcers developing on the inside of his cheeks; his gums were swollen and bled if he pressed on them, and the interior of his right nostril was purple and sore. By early autumn, he was faint and vomiting, and spent one memorable night passing—and taking notes on—14 watery stools. A less committed researcher might have been discouraged if his gums turned black, purple streaks showed up on his shoulder, and a "disagreeable, fetid, yellowish fluid" built up in his mouth, but Stark took only a brief break to allow his gums to heal before plunging back in. This time, in addition to bread and water, he began experimenting with what he'd decided were his highest priority foods. Olive oil. Milk. Boiled beef. Roast goose.

Eventually, after a bad experience with "puddings"—thick concoctions of flour and salted water mixed with oils, beef fat, butter and suet—Stark decided to investigate what, by this point, he considered the logical next question: what would happen if he left out the flour? The result was a diet of water, suet, egg yolks, and jellied calves' feet. It proved too disgusting even for him.

"The mess was so disagreeable, that, after tasting it, though I was extremely hungry, I could not eat it," he wrote, "and therefore dined on one pound of figs." It was a notable aberration, for along with a pint of currants consumed in honour of Boxing Day (and some plums, which he didn't record eating but whose skins appeared mysteriously in his stools), those figs were the only fruit or vegetable—in the entire course of experiment—that Stark mentioned having consumed. By the end of the year, after concluding that his pudding diet had left him "not so lively as usual," he did consider testing the effects that other fresh fruits and vegetables might have on his health. But instead, he decided to study honey pudding and Cheshire cheese.

On December 31, Stark's gums were so painful that it was "troublesome … to eat even the crumb of bread." On February 18, his bowels felt "as if they had been bruised." On February 23, 1770, eight and a half months into his nutritional adventures, feverish, bed-ridden and eating a diet of bread and rosemary tea, William Stark died. He was 29 years old.

Anyone reading Stark's story today can probably guess what the problem was: his monotonous diet was extremely low in vitamin C, an essential micronutrient that's prevalent in fruits, vegetables, and many other foods that Stark was not consuming. The poor man was therefore likely killed by the vitamin C deficiency disease of scurvy, which he could have prevented (and cured) simply by eating fruit and vegetables. *Choose the figs, William!* I wanted to tell him. *Not the cheese. The figs!*

But unfortunately, Stark didn't know about vitamins. In his time, no one did. It was only in the late-nineteenth century—roughly 100 years after Stark's death—that scientists began to suspect a connection between diseases like scurvy and chemicals in food. It took decades for the concept of vitamins and deficiency diseases to gain widespread acceptance, and years more before any vitamin was actually discovered. The word's first print appearance was only in 1912.

We've since become convinced that vitamins are always good for us—and that the more we can get, the better. In reality, however, most of us know nearly nothing about vitamins. And while our easy access to vitamins means that we're unlikely to meet the same fate as William Stark, our faith in them—combined with the way they've shaped our outlook on nutrition—is doing us a different sort of harm. It's easy to look back at Stark with pity. But in many ways, despite all that we have since learned about vitamins and nutrition, we remain similarly naïve.

Today, we're such believers in vitamins' inherent goodness that we don't realize just how much scientists still don't understand about how vitamins work in our bodies, or how much of each we each require. We're not aware that vitamins (and our enthusiasm for them) are what opened the door for the rotating cast of supposed wonder-nutrients that intrigue and confuse us today, whether they be probiotics or antioxidants or omega-3s. We don't notice the ways that food marketers and dietary supplement makers use synthetic vitamins to add a veneer of health to otherwise unhealthy products; nor do we acknowledge the extent to which we use vitamins and these other vitamin-inspired nutrients to give ourselves a free pass to overeat foods of all kinds. And we certainly don't recognize the irony of our vitamin obsession: encouraging the idea that dietary chemicals hold the keys to good health is, in fact, making us less healthy.

One assumption about vitamins is definitely true: we do indeed require them. The thirteen dietary chemicals that we call vitamins affect each one of us every minute of every day, helping us to think and speak and move our muscles, pull energy from what we eat, even see the words on this page. Deficiencies in these chemical compounds have killed and continue to kill millions of people around the world, and when administered soon enough, vitamins can be astoundingly powerful, even miraculous—give vitamin A to a child suffering from the vitamin A deficiency condition of night blindness, and she can regain full sight within days. Our need for them is ultimately no more avoidable than our need for air; while vitamins are normally invisible, their implications are profound.

But the very power of vitamins makes them a double-edged sword. Their ability to save lives has promulgated the idea that they can perform miracles in all of us, regardless of whether we're actually deficient; this, in turn, has led to beliefs in vitamins that are based more on faith than fact. When we seek out vitamins today, it's not because we're worried about night blindness, or pellagra, or beriberi, or any of the other conditions that vitamins can actually prevent and cure—true vitamin deficiency diseases have become so uncommon in the developed world that most of us don't even know their names.

Instead, we use vitamins as insurance policies against whatever else we might (or might not) be eating, as if by atoning for our other nutritional sins, vitamins can save us from ourselves. We think that vitamins will help us live longer and stay healthier, even prevent or reverse disease. Perhaps that's why when we hear the word "vitamin," our minds often jump immediately to pills, turning substances found naturally in foods into something we don't just "eat," but "take." Yet while we all know that medicines can have side effects, and that no one drug could possibly solve all our problems, we assume that vitamins are both panaceas and entirely risk-free.

In a way, our attraction to vitamins—and general interest in nutrition—is perfectly logical: no one wants to be sick. But that doesn't explain how the term "vitamin," a word coined by a Polish biochemist before any vitamin had been chemically identified, has come to be synonymous with health. Isn't it odd, for example, that cyanocobalamin and alpha-tocopherol sound intimidating, even sinister, while vitamins B12 and E—which are names for the same substances—seem incontrovertibly good? Isn't it strange that we worry about hydrogenated oils, high-fructose corn syrup, artificial sweeteners, and GMOs, but allow synthetic vitamins to be added to nearly anything—and then use the very presence of those vitamins to define the food as healthy? How can we simultaneously harbour distrust toward drug makers, but accept extravagant claims on foods and supplements, whose manufacturers make billions of dollars from us each year, at face value? If we were to ask these questions, we might reach an uncomfortable conclusion: that both individually and as a society, we have been seduced by a word.

Despite its influence on our daily lives, most of us are unaware of this seduction. I know this from personal experience—I'm a journalist with a particular interest in health and nutrition, and I also have type 1 diabetes, an incurable autoimmune disease that forces me to pay constant attention to how what I eat will affect my body. I inspect Nutrition Facts panels whenever I encounter them, and I follow stories about nutrition the way that other people follow sports. For reasons both personal and professional, my life depends on being knowledgeable and thoughtful about food.

As a result, until recently, I thought I understood vitamins. I could tell you that they were essential substances that we need to get from our diets, and like anyone who paid attention in school, I was aware that sailors used to suffer from something called scurvy. But I didn't really understand why animals and plants need vitamins, how they were discovered, or even the technical definition of what a vitamin actually is. Instead, like many people, I just aimed for 100 percent of my daily requirements, ate a lot of kale, popped a multivitamin when I remembered it, and bought into the common assumption that the more vitamins a food contained, the better for me it must be.

Part of my lack of curiosity stemmed from my assumption that vitamins represented a problem that had already been solved, and which I therefore didn't need to worry about myself. So when I discovered how much scientific uncertainty still surrounds vitamins (not to mention that billions of people in the developing world still don't have sufficient access to them), I was both shocked and shaken. If some of the most basic questions about vitamins still have no answers, then what else don't we know about nutrition? And how should this affect the way we think about food?

In addition to defining the "what" of vitamins, I wanted to understand where the vitamins in our foods came from and how they got there. I wanted to figure out—as a consumer, not a chemist—how to wade through the hype surrounding vitamins, and how the history of vitamins could help me make better nutritional choices today. I wanted to consider a potentially humbling question: barely a hundred years ago, the very concept of a vitamin was brand new. So what facts about nutrition might our grandchildren be amazed to learn that we don't know? And most importantly, I wanted to explore our often unacknowledged psychological and emotional relationship with vitamins, a state of mind so fraught with fear and hope, belief and blind acceptance, that it could be called a complex.

And so I read memoirs and histories, talked to scientists, and went on field trips. I had my blood analyzed and my genes tested; I ate military rations and alfalfa tablets, yeast cakes, nutrition bars, and caffeinated meat. The tale that emerged is one of insane asylums and conspiracies, poison squads and political maneuvering, irradiated sheep grease, smuggled rats, even a doctor so intent on proving his theories about nutrition that he injected himself with his patients' blood. I found answers to many of my questions, but I also uncovered unexpected controversies, continued uncertainties, and evidence of how our vitamin complex, as I've come to call it, affects our daily lives—and our health—in far more ways than we might think.

This book is not an instruction manual, a straight history, or an in-depth analysis of every vitamin. Nor is it an attack on man-made synthetic vitamins themselves, which continue to prevent deadly deficiencies all over the world. Rather, it's an attempt to use the story of vitamins to explore our bizarre relationship with nutrition, and propose an alternative to our obsessive quest for perfection--one that's focused on acting on what we do know, instead of being tormented by what we don't.

It starts with the astonishing story of where our vitamins actually come from, and how dependent on synthetic vitamins we are. After exploring the discovery of vitamins and how they went from being a scientific breakthrough to a public craze, it looks at some of the current devastating consequences of vitamin deficiencies in the developing world and asks why our food supply contains so many man-made vitamins to begin with. It examines how vitamins primed us to accept (and defend!) the much broader category of dietary supplements: more than 85,000 products, which go way beyond standard multivitamins, that are not required to be tested for safety or for efficacy before being sold. It investigates the limits of vitamins' powers, whether megadoses (or even multivitamins) are beneficial, and why our personal beliefs in vitamins and the recommendations of official health organizations are often so contradictory. And it ends with a glimpse toward the future of nutrition, exploring several current questions about how food interacts with our bodies— questions whose answers may eventually be considered just as obvious to our descendents as vitamins are to us today.

Ultimately, the story of our vitamin complex reflects our abhorrence of uncertainty—a fundamental discomfort that leaves us hungry for solutions and susceptible to fads. This is especially true in the case of nutrition, a subject that we want to have clear-cut answers, but which turns out to contain far more nuances and unanswered questions than we like to admit. But the story of our vitamin complex also proposes an alternative: what if, instead of running from this uncertainty, we embraced it?

If we did so, we might find ourselves freed from the anxiety—and craziness—that defines our modern approach to nutrition. We might gain a sense of confidence about how to interpret the onslaught of news stories and food advertisements and conflicting dietary advice that bombards us each day. And we might rediscover something both surprising and empowering: that while nutrition itself is amazingly complex, the healthiest, most scientific, and most pleasurable way to eat is not that complicated at all.

1

High Seas and Hi-C

[W]hat is the function of these vitamines? If fats
and carbohydates provide the fuel, and proteins the material
for tissue supply, and mineral salts are needed for bone construction,
etc., just what do the vitamines supply? We do not know.
—BENJAMIN HARROW, *The Vitamines: Essential Food Factors,* 1922

The first time I saw a vitamin in pure form—as opposed to just gulping one down in a pill—was in Parsippany, New Jersey. It was a drizzly November day, and I was visiting the Nutrition Innovation Center, a product-development facility run by the world's largest synthetic vitamin producer, the Dutch company DSM.

Companies come to the centre to brainstorm and create new products, harnessing the expertise of DSM's chemists and flavour technicians to add vitamins and other so-called functional ingredients to their foods. But I hadn't come to develop a new fortified beverage or cereal or snack bar. My goal was more basic: after more than three decades of eating and taking vitamins, I had come to the centre to learn what vitamins actually *are.*

My host for the day was DSM's senior director of global technical marketing, a French-born pharmacist and PhD named Jean-Claude Tritsch, who had ear-length greying hair and wore a pink V-neck sweater. We were in the room where product concepts are shared and sampled with food and supplement companies, and Tritsch was explaining the basics of vitamins from behind a wet bar as I sat perched on a high stool at a granite countertop, a selection of product prototypes arranged in front of me.

When we hear the word "vitamin," many of us immediately think of pills; we also tend to mistakenly apply the term to *all* dietary supplements, and often lump vitamins and minerals together. But as Tritsch explained, there are actually only thirteen human vitamins, all of which are organic compounds that occur naturally in food. Four are fat-soluble, meaning they dissolve in fat and need fat to be absorbed: A (retinol), D (cholecalciferol), E (tocopherol), and K (phylloquinone). The other nine are water-soluble: C (ascorbic acid) and the eight substances grouped together in what's called the B complex— B1 (thiamin), B2 (riboflavin), B3 (niacin), B5 (pantothenic acid), B6 (pyridoxine), B7 (biotin, also sometimes referred to as vitamin H), B9 (folate/folic acid), and B12 (cobalamin). Sometimes choline is counted as a fourteenth vitamin, but usually the roster ends at thirteen. (Some vitamins come in more than one chemical form—the parentheticals refer to the most common or the most relevant.)

Unlike the macronutrients (fat, protein, and carbohydrate), vitamins are not burned as fuel; instead, their primary role is to facilitate chemical reactions in our bodies that keep us alive. Vitamins, Tritsch told me, are thus considered essential micronutrients—*essential* because our bodies require them but can't make sufficient quantities, which means we need to get them from outside sources, and *micro* because we only need them in really small amounts, typically fewer than 100 milligrams a day.

Indeed, we need vitamins in amounts so tiny that it's difficult to visualize them, let alone to believe that our lives depend on them. The amount of folic acid that pregnant women are told to take to prevent devastating neurological defects in their babies is 240 *micro*grams a day, less than the weight of two grains of table salt. The Recommended Dietary Allowance for vitamin D, without which you won't be able to properly absorb calcium and your bones will soften, is 15 micrograms (600 IU), one-sixteenth of that for folic acid. And the RDA for B12, a vitamin whose deficiency can cause depression, delusions, memory loss, incontinence, nerve damage, and in extreme cases life-threatening anemia, is smaller still, just 2.4 micrograms—0.0000024 grams. That's 1/100th of the weight of the requirement for folic acid, the equivalent of 1/67th of one grain of salt.

Searching for a way to make those tiny numbers tangible, Tritsch let me taste and smell several samples of pure vitamins that were kept on hand at the lab. Vitamin C was a talc-like white powder, tart like a sherbet lemon and very irritating, I discovered with the help of a paper cut, if rubbed into an open wound. Thiamin was bitter and white. Powdered riboflavin was the colour of butternut squash. Folic acid was yellow and tasted chalky. A and D were clear, sticky, meltable crystals, so concentrated and unstable that they're usually dissolved in oil. E was a tasteless, viscous clear fluid. Vitamin B12 was bright pink.

By the time I left the Innovation Center, I'd seen diagrams of vitamins' chemical structures and magnified photographs of individual molecules, colourful crystals that sparkled in the light. But even after I'd touched them, tasted them, and smelled them, I still couldn't wrap my head around what I was experiencing. It seemed somehow impossible that these odourless, unassuming substances could be essential for keeping me—and every one of us—alive.

The problem, I realized, was that I still didn't understand what vitamins do in our bodies—which is a necessary concept to grasp if you want to understand why a deficiency could kill you. So I decided to look for an explanation in the vitamin I thought I knew the best: vitamin C.

Most people know that if you don't have enough vitamin C, you'll develop a vitamin deficiency disease called scurvy, and you have probably heard tales of sailors on long sea voyages whose teeth fell out as a result. But having loose teeth, while certainly unpleasant, doesn't sound life-threatening. And besides, scurvy can be cured by drinking orange juice. How serious could it really be?

Really serious, it turns out. Far from just affecting their gums, scurvy killed more than two million sailors between Columbus's 1492 transatlantic voyage and the rise of steam engines in the mid-nineteenth century. It was such a problem that ship owners and governments counted on a 50 percent death rate from scurvy for their sailors on any major voyage; according to historian Stephen Bown, scurvy was responsible for more deaths at sea than storms, shipwrecks, combat, and all other diseases combined.

Scurvy starts with lethargy so intense that people once believed laziness was a cause, rather than a symptom, of the disease. Your body feels weak. Your joints ache. Your arms and legs swell, and your skin bruises at the slightest touch. As the disease progresses, your gums become spongy and your breath fetid; your teeth loosen and internal hemorrhaging makes splotches on your skin. Old wounds open; mucous membranes bleed. Left untreated, you will die, likely as the result of a sudden hemorrhage near your heart or brain.

Bown quotes a survival story written by an unknown surgeon on a sixteenth-century English voyage that illustrates scurvy's horror. "It rotted all my gums, which gave out a black and putrid blood," he wrote. "My thighs and lower legs were black and gangrenous, and I was forced to use my knife each day to cut into the flesh in order to release this black and foul blood. I also used my knife on my gums, which were livid and growing over my teeth. . . . When I had cut away this dead flesh and caused much black blood to flow, I rinsed my mouth and teeth with my urine, rubbing them very hard. . . . And the unfortunate thing was that I could not eat, desiring more to swallow than to chew. . . . Many of our people died of it every day, and we saw bodies thrown into the sea constantly, three or four at a time."

Scurvy affected many of the explorers we learned about in grade school—Vasco da Gama lost his brother to it; Ferdinand Magellan watched it kill many of his men, who had been reduced, he wrote, to existing on "old biscuit reduced to powder, and full of grubs, and stinking from the dirt which the rats had made on it when eating the good biscuit." Scurvy killed so many men on the 1740–1744 voyage commanded by a British captain named George Anson that it is considered one of history's worst medical disasters at sea.

When reading about such experiences, it's difficult not to want to travel back in time, grab these men by the shoulders, and beg them to eat some lemons. The idea that certain foods can cure scurvy wouldn't even have been a new idea—in 1535, French explorer Jacques Cartier reported that after his ships had become frozen in the St. Lawrence River, his men were saved from scurvy by a special tea, prepared by the local Native Americans from the bark and leaves of a particular tree. In the 1500s and 1600s, several ships' captains suggested that there might be a connection between produce and scurvy. In 1734, a Dutch physician named Johannes Bachstrom came up with the term "antiscorbutic"—against scurvy—and used it to describe fresh vegetables.

Even Anson—captain of the aforementioned disastrous voyage—made a point of loading up on oranges whenever possible, and his chaplain, Richard Walter, described certain vegetables as being "esteemed to be particularly adapted to the cure of those scorbutic disorders which are contracted by salt diet and long voyages." But while many mariners recognized that there was a connection between sailors' diets and their susceptibility, no one knew the true cause of scurvy, or what made certain foods antiscorbutic.

Today, scientists understand the connection, and it has to do with what vitamins are actually doing in our bodies. Despite their chemical differences, all vitamins play crucial roles in our metabolism, a term that refers to the series of chemical reactions that occur in our cells. Though we are rarely aware of these metabolic chemical reactions, our lives depend on them. Walking down the street requires them. Reading a book requires them. So does forming scar tissue, developing a baby, or creating any type of new cell. Chemical reactions build and break down muscle, regulate body temperature, filter toxins, excrete waste, support our immune systems, and affect (or indeed cause) our moods. They generate the energy we need in order to breathe, and use the oxygen that we breathe to pull energy from food. They allow us to feel and see and taste and touch and hear. Our metabolisms aren't just a facet of our lives—they *are* our lives. Without these metabolic chemical reactions, we would be as inert and inanimate as stone.

The problem with many of these reactions, however, is that they're way too slow—if they were left to run at their own speed, life would grind to a halt. Our bodies get around this issue with the help of enzymes, which are large protein molecules that kick-start and speed up specific chemical reactions, often making them occur millions of times faster than they would on their own. But our bodies sometimes need help making enzymes, and enzymes sometimes need help doing their jobs. That's where vitamins come in: two of their primary functions are to help our bodies create enzymes and to aid enzymes in their work. While enzymes speed up chemical reactions without being destroyed, most of the chemical reactions that depend on vitamins actually use up the vitamins. That's why we need a continuous external supply.

It makes sense, then, that vitamin deficiencies cause problems, because without adequate vitamins, every enzymatic process that depends on those vitamins will come screeching to a stop. In the case of scurvy, the issue is collagen, a primary structural protein in our muscles, skin, bones, blood vessels, cartilage, scars, and other connective tissues that makes up some 30 percent of the protein in the human body. Collagen holds our tissues together; the word itself is derived from the Greek word for "glue." Without collagen, our bodies would come apart from within—hence the hemorrhaging, broken bones, and loose teeth of scurvy. We make collagen from its precursor, procollagen, with the help of enzymes. But those enzymatic reactions can't happen—and thus collagen cannot be formed—without vitamin C.

With that said, scientists still don't fully understand all the nuances of what vitamins do in our bodies, how they do it, or what the long-term effects of moderate deficiencies might be. That, in turn, makes it extremely difficult to create precise nutritional recommendations. In the words of a 2003 report from the nongovernmental Food and Nutrition Board at the National Academy of Sciences' Institute of Medicine, "[s]cientific data have not identified an optimum level for any nutrient for any life stage or gender group, and [today's nutritional recommendations] are not presented as such." Instead, the same report explains that "a continuum of benefits may be ascribed to various levels of intake of the same nutrient."

In fact, the RDAs themselves—which many of us use as personalized scorecards for our diets—are actually not meant to be personal at all. Instead, they're designed to meet the nutritional needs of 97 to 98 percent of all people, which means that the majority of us could get by just fine on less. (There's also no need to get 100 percent of your RDA every day—what's important is your consumption over time, since our bodies maintain stores of most micronutrients.) And even with that generous built-in margin for error, the Food and Nutrition Board, which is responsible for updating the RDAs in the United States, still has not established adult RDAs for biotin, pantothenic acid, or vitamin K, and there are no RDAs for infants up to one year old for *any* vitamin.

It's also still surprisingly difficult to measure vitamins, whether in our bodies or in foods. Blood tests exist for several, but there are often problems with standardization (that is, results from the same sample can vary from one lab to the next), and there's continued controversy over what the cutoff for "deficiency" should be. Adding to the challenge, some vitamins are stored in inaccessible places in the body—the most accurate way to measure vitamin A would be a liver biopsy—and our vitamin levels can vary considerably by day or by season depending on what we eat. If you eat a lot of pink grapefruit, for example, your vitamin C level will spike within hours. If you smoke a cigarette, it will drop (as will that of folate). If it's summertime, your vitamin D level will likely be higher than it is in the winter, when you're less likely to be out in the sun and usually cover more of your skin with clothing. And as if that's not enough, the vitamin information on food labels is often based on composites, meaning that even if you knew your body's precise vitamin requirements, you wouldn't be able to calculate exactly what percentage of those requirements were represented by the food on your plate.

But despite these continued uncertainties, we definitely know more than early explorers, who weren't aware of vitamins at all. As for the era's doctors and scientific thinkers, they not only lacked the analytical tools and chemical knowledge necessary to even conceive of a nutritional deficiency disease, but many popular hypotheses about scurvy's cause were still related to the ancient theory of the humours, which assumed that people's innate constitutions influenced their likelihood of getting sick, and that disease should be treated by balancing four "humours" that flowed through the body: black bile, yellow bile, blood, and phlegm. Supposed triggers were even more haphazard. According to author Frances Rachel Frankenburg, they ranged from fatigue and depression to homesickness, contagion, seawater, damp air, copper pans, tobacco, hot climate, cold climate, rats, heredity, contagion, fresh fruit (whoops), too much exercise, too little exercise, sea air, salted meat, poor morals, and filth.

And even if the concept of vitamins had been familiar, vitamin C would have been a tough one to figure out. Humans and several other simians—along with guinea pigs and fruit bats—are the only mammals that can't make their own vitamin C. In other creatures, it's referred to as "ascorbic acid" (shorthand for antiscorbutic) and, since their bodies can produce it in sufficient quantities, isn't considered a vitamin at all.

It's also not obvious where to find vitamin C. There are large amounts in liver and kidneys, but not in muscle meat. Eggs and cheese don't have any. Cabbage and broccoli have a lot. A half cup of pears will give a woman about 4 percent of her 75 mg/day RDA, but the same amount of kiwifruit will give her 111 percent. Once the connection between citrus fruit and scurvy had been recognized and accepted, Britain often supplied its sailors with limes—which it chose instead of lemons because it controlled colonies that grew them (hence the nickname "limey" for British sailors). But this thriftiness came at a price: limes have only half as much vitamin C as lemons and oranges. Preparation matters, too. The proponents of "rob," a popular treatment made from boiled-down citrus juice, had the right idea, except guess what? Vitamin C is destroyed by heat—not to mention cutting, bruising, exposure to air, and being cooked in copper pots.

As a result, the confusion over scurvy was so great that even James Lind, the person who gets the most credit for establishing that citrus fruit cures scurvy, overlooked his own discovery—making vitamin C an early example of how complicated the overall process of discovering vitamins turned out to be.

Lind was a Scottish physician who served as a naval surgeon on HMS *Salisbury* in 1747, and devised what is considered to be one of the world's first controlled experiments. First, he took twelve sailors who were sick with scurvy and divided them into six pairs. All the men ate the same food and lived in the same quarters on the ship; the only difference was their treatment. Lind gave each pair daily doses of one of six different supposed scurvy cures: a quart of hard cider, twenty-five drops of vitriol (a mixture of sulphuric acid and alcohol), two spoonfuls of vinegar, a half pint of seawater, two oranges and one lemon, and last, an "electuary"—a creative mix of garlic, mustard seed, balsam of Peru, dried radish root, and gum myrrh, shaped into a pasty concoction the size of a nutmeg. Lest that treatment not sound random enough, those sailors also got barley water treated with tamarinds and an occasional laxative dose of cream of tartar. With the exception of the citrus fruit, which ran out in less than a week, Lind administered the treatments for fourteen days.

As the diversity of treatments indicates, Lind's experiment had no foregone conclusion. Nonetheless, it didn't take long for one intervention to emerge as better than the others: the men treated with citrus fruits recovered so thoroughly and rapidly that they were able to help Lind care for the others. Because of this experiment, Lind is often given historical credit for recognizing citrus as a definitive cure for scurvy. But that's not actually what happened.

Instead, when Lind retired from the navy in 1748, he got to work on the first edition of a massive book called *A Treatise of the Scurvy: Containing an Inquiry into the Nature, Causes, and Cure, of That Disease Together with a Critical and Chronological View of What Has Been Published on the Subject.* True to its sweeping title, it ended up being some four hundred pages long. Lind described his crucial experiment in five paragraphs about two hundred pages into the book, and condensed the key result into one seriously downplayed sentence: "As I shall have occasion elsewhere to take notice of the effects of other medicines in this disease, I shall here only observe that the results of all my experiments was, that oranges and lemons were the most effectual remedies for this distemper at sea."

Lind wasn't trying to bury the lead; he just didn't recognize the significance of his results. Sure, the oranges and lemons had cured scurvy, but the sailors who got the cider seemed a little better, too. This is plausible, since the unrefined hard cider Lind distributed might have contained a little of the vitamin. And so rather than dwell on citrus, Lind moved on to describe his own humours-inspired explanation of scurvy: it was actually a digestive disease caused by blocked sweat glands.

By the time Lind published the third edition of his book in 1772, he had completely lost sight of what we now consider his most important observations. While he did still think lemon juice might be effective against scurvy—he thought it might clear out those blocked sweat glands, especially if mixed with wine and sugar—he included so many disclaimers that his argument was hardly convincing. "I do not mean to say that lemon juice and wine are the only remedy for the scurvy," he wrote. "This disease, like many others, may be cured by medicines of very different, and opposite qualities to each other, and to that of lemons."

Nonetheless, progress was gradually made. It had to be: as the size of the world's navies increased, the problem of scurvy only grew worse—and it wasn't long before the search for a cure for scurvy became what Stephen Bown describes as "a vital factor determining the destiny of nations." In 1795, a physician named Gilbert Blane convinced the British navy to issue some form of lemon juice to its sailors. His order likely changed the course of history when it helped Great Britain to successfully defend itself from a Napoleon-led invasion by setting up a blockade of the English Channel. This blockade, during which many ships spent months on the water without coming to port, went on for twenty years—a feat that scurvy would never have allowed.

Yet no matter how many times the connection between scurvy and produce was demonstrated, people kept forgetting it; cures for scurvy—like those for many of the other vitamin deficiency diseases—continued to be lost and found and lost again. Scurvy appeared in Arctic explorations of the 1820s and the 1848–1850 American gold rush. Florence Nightingale reported entire shiploads of cabbage being tossed overboard during the Crimean War of 1853–1856 at the same time that soldiers were perishing from the disease. (The cabbage had been sent specifically to treat scurvy, but thanks to bureaucratic snafus, no one had ordered it to be distributed in the men's rations.) Scurvy plagued prisoner-of-war camps in the twentieth century, and even emerged among the babies of wealthy and educated Americans and Europeans in the late 1800s and early 1900s, thanks to unfortified pasteurized cow's

milk (the heat destroyed the vitamin C). Nearly a century would pass after
the British blockade before anyone truly understood *why* fresh fruit or cabbage was effective in preventing scurvy; till then, it continued to reappear
wherever diets and circumstances allowed.

Though it might seem strange to us today, scurvy was in its time a very modern disease—an example, among many others in the story of nutrition, of how
advancements in one area can lead to problems in another. True, scurvy
existed in ancient times and was common in Northern Europe during the
Middle Ages, when harvests were too small to provide adequate vitamin C
through the long winter. But for seafarers, technology is what truly made it a
concern: it only became prevalent after the development of long-distance
ships, navigational techniques that freed them from the shorelines for months
at a time, and rations that, while often dangerously low in multiple vitamins,
had enough calories to ensure that sailors wouldn't starve.

In a way, scurvy was therefore an early example of a disease of civilization,
a category of ailments caused by human-driven changes in the environment.
Just as public health experts now worry about what the rising rates of coronary
heart disease and type 2 diabetes will do to our long-term productivity, their
predecessors had the same concerns about scurvy. Despite more than a century of separation, the underlying concerns are the same.

And even though we now know how to prevent them, vitamin deficiency
diseases will never truly be relegated to the past. An estimated two billion
people currently don't have access to adequate vitamins. At least four outbreaks of scurvy have been reported worldwide since 1994. The bone-
softening vitamin D deficiency disease of rickets is prevalent in Indian slums
and other areas in the developing world, and, while rare, cases have even been
reported in British and American children whose diets and lifestyles don't
provide them with adequate amounts of vitamin D. Millions of people, particularly children, are deficient in vitamin A, and will go blind or die as a result.
Folic acid deficiencies continue to cause devastating birth defects. General
vitamin deficiencies can and do occur in refugee camps, prisons, and in any
place—or in any population—without access to nutritionally adequate food.

The reason is simple, if strange to think about: despite the steady march
of scientific advancement that separates us from our predecessors, there is

nothing about our modern bodies that makes us invulnerable to scurvy or any other vitamin deficiency. Human beings have evolved to need vitamins; our bodies can't function without a continuous supply. Unlike infectious diseases, which can be prevented and cured with vaccines and drugs, and sometimes wiped out entirely, there is no way the threat of vitamin deficiency diseases can ever be eradicated, or the diseases themselves permanently "beaten." Instead, consistently good nutrition is their only prevention and their only cure. Today in the West, scurvy might seem as distant as the Black Death. But take away our oranges, or our fortified foods, or our pills, and we'd be just as vulnerable as those sorry sailors.

2

Plants and Plants

The discovery that tables may groan with food and that we
may nevertheless face a form of starvation has driven home
the fact that we have applied science and technology
none too wisely in the preparation of food.
—*New York Times* EDITORIAL, 1941

n 2011, the *Journal of Nutrition* published a report with shocking implications. Titled "Foods, Fortificants, and Supplements: Where Do Americans Get Their Nutrients?" it found that "large percentages of [the intakes of] vitamins A, B6, B12, C, and D as well as thiamin, riboflavin, niacin, folate, and iron were from fortification and/or enrichment" with synthetic vitamins.[1] Without supplements and enriched and fortified products, the authors estimated that 100 percent of us would fail to meet the Estimated Average Requirement for vitamin D, 74 percent for vitamin A, 46 percent for vitamin C, 93 percent for vitamin E, 51 percent for thiamin, 22 percent for vitamin B6, and 88 percent for folate. To put it differently—despite our wealth, despite our sophisticated food supply—if it weren't for synthetic vitamins, we would be at risk of serious vitamin deficiencies.

Those figures are especially astounding given that EARs, which represent the amount of a nutrient thought to meet the needs of 50 percent of the people in a particular age group, are far lower than the 97–98 percent target of the RDAs. What's more, even *with* supplements, fortification, and enrichment, the study found that "considerable percentages" of Americans had

---

[1] In general, enrichment means replacing the micronutrients—like vitamins and minerals—that were naturally present in the food but which processing removed or destroyed; fortification refers to adding extra or new micronutrients beyond what the food originally contained.

intakes below the EARs in vitamins A, C, D, and K, as well as calcium and magnesium.

The paper's findings are a damning indictment of our diets, especially when you consider that most of us are consuming far more calories than we require. But they also raise the question of where our vitamins—both synthetic and natural—come from to begin with. It's an issue worth exploring, because it turns out that there are enormous discrepancies between where we *think* we're getting our vitamins and what the realities of our food supply actually are. And these discrepancies point toward a conclusion that is fundamentally disturbing: we are far less nutritionally self-sufficient than we assume.

Let's start by taking a look at the origins of the vitamins found naturally in plant and animal foods like liver or kale. Many animals can make the chemicals we call vitamins and therefore don't need to get them from an external source. The other vitamins in animal products come either from vitamins that naturally occur in the foods the animals consume or from synthetic vitamins added to their feed—in fact, about 50 percent of the world's supply of synthetic vitamins is used for this purpose. Regardless of their source, these vitamins are likely doing the same things in animals' bodies, like facilitating enzymatic reactions, as they are in our own.

As for plants, their vitamins are entirely self-made: while not every plant makes or needs every human vitamin, in combination, plants naturally make all of the human vitamins except for D, B12, and A. (Plants make beta-carotene, which our bodies can turn into vitamin A, but they don't make vitamin A itself; some fungi, like mushrooms, can create vitamin D if they are exposed to ultraviolet light, but that usually only occurs if humans deliberately intervene.) Just as our bodies need vitamins to facilitate reactions that convert food into energy, plants need vitamins for photosynthesis—the miraculous process of creating sugar and starches from sunlight and carbon dioxide. Without vitamins, this process couldn't occur.

And just like our metabolism, photosynthesis is a sloppy process, producing all sorts of potentially damaging molecules called free radicals that are created when light breaks down water. One of vitamins' roles in plants—as well as in humans—is to act as antioxidants, molecules that are able to neutralize these free radicals so that they don't cause harm. This means that the more photosynthesis a plant engages in—whether because it's located in a particularly sunny location or because its natural pigmentation gathers more light—the higher the levels of vitamins and other antioxidant chemicals that it's likely to have.

The fact that darker colours absorb more light—which is accompanied by higher levels of potentially damaging radiation—is one reason why light-coloured vegetables like iceberg lettuce tend to be lower in vitamins and other micronutrients than dark leafy greens such as kale, spinach, and spring greens. Fruits and vegetables can also have different nutrient profiles depending on their variety (a Granny Smith has a different sugar and vitamin content than a Red Delicious apple, for example), where they were grown, when they were picked, how much sun and water they got, how they were processed, and how long they've been in storage. There can even be differences in the concentration of nutrients in one single piece of produce. The stem end of a fruit can contain a higher concentration of vitamins than the base; the outer leaves frequently contain more vitamins than the inner; and yes, it's true: vitamins are often concentrated in fruits' and vegetables' skins.

Our distant ancestors were likely able to make some of the chemicals we now consider vitamins, but they lost the ability. This might be because they outsourced the job to plants. For example, humans have the genes necessary for producing vitamin C, but our version contains a disabling mutation that prevents us from actually doing so. Researchers hypothesize that, much like a muscle that atrophies with disuse, this mutation may have developed as a result of the more than ample supply of vitamin C that was available from fruits and other plants. But regardless of the original reason, the result is that we can no longer make our own.

Today, photographs in health magazines often make it seem like we're getting our vitamins from walnuts and blueberries—and it is true that when the first vitamin supplement products hit the market in the 1920s and 1930s, many of the products were extracts of natural sources. To get cod-liver oil, for example, you floated cod livers in warm water till the vitamin-rich oil rose to the surface and then skimmed it off; vitamin C often came from rose hips. But today, while it's possible to extract some vitamins from food—like vitamin E from soybeans—it's often prohibitively expensive to do so. It can also be environmentally destructive, because extraction requires chemical solvents, some of which can be toxic.

"It's not just pressing a mango or an apple to get juice," Jean-Claude Tritsch at DSM explained.

And given the tiny quantities of vitamins most foods contain, extraction from natural sources is also not practical. It would be impossible, for example, to satisfy the world's vitamin C demands via real oranges or lemons, given that literally tons of vitamin C are used in a nonnutritive role as antioxidant preservatives. As a result, the majority of the vitamins we consume today come from a completely different type of plant.

That is to say, a factory; most vitamins in supplements or enriched or fortified foods are synthetic, man-made substances that have been formulated into premixes that companies can add to their products. Many are produced by chemical reactions that use a catalyst like heat, acid, or pressure to rearrange the molecular structure of two or more chemicals into a vitamin; others are derived from biotechnological techniques, which usually means finding (or genetically engineering) microbes that can manufacture vitamins for you. Vitamin B12 in particular is so molecularly complicated that its production is nearly always outsourced to bacteria.

As for your synthetic vitamins' raw ingredients, they're definitely not rose hips. Instead, here is how journalist Melanie Warner describes the production of vitamin C, which is often referred to as being corn-derived, in her book about the American food industry, *Pandora's Lunchbox*:

> It starts not with corn kernels or even corn starch, but sorbitol, a sugar alcohol found naturally in fruit and made commercially by cleaving apart and rearranging corn molecules with enzymes and a hydrogenation process. Once you have sorbitol, fermentation starts, a process that tends to muck up surrounding air less than chemical synthesis (although it's been known to cause problems with water pollution). The fermentation is done with bacteria, which enable more molecular rearrangement, turning sorbitol into sorbose. Then another fermentation step, this one usually with a genetically modified bacteria, turns sorbose into something called 2-ketogluconic acid. After that, 2-ketogluconic acid is treated with hydrochloric acid to form crude ascorbic acid. Once this is filtered,

purified and milled into a fine white power it's ready to be shipped off as finished ascorbic acid, mixed with other nutrients and added to your Corn Flakes.

Warner points out that, as complicated as this process might be, vitamin C "is about as food-based as vitamins get." According to Warner, synthetic vitamin A's raw ingredients include acetone and formaldehyde; niacin is often made using a waste product of something called nylon 6,6—a synthetic fibre that's often used for commercial carpets, airbags, zip ties, and conveyor belts. Thiamin is synthesized from chemicals derived from coal tar.

If that sounds strange, consider the source of most of the world's synthetic vitamin D: sheep! Or, more specifically, lanolin, the greasy substance found in wool. Lanolin is used in many products, including cosmetics, moisturizers, and industrial lubricants (it inhibits rust)—but it can also be chemically purified and irradiated to produce cholecalciferol, the form of vitamin D that our bodies produce in response to sunlight. This means that most of the vitamin D you consume in capsules, milk, cereals, and other fortified foods comes from the same source as your favourite sweater.

It's important to note that there's nothing inherently wrong, healthwise, with sheep-derived vitamin D or any of the other raw ingredients used for vitamins, odd or industrial though they may sound. In most cases, the resulting synthetic vitamins are chemically identical to the forms found in nature, which means that our bodies use them in exactly the same ways.[2] The primary reason most nutritionists recommend getting your vitamins from food rather than supplements is not that synthetic vitamins are bad, but that natural foods contain countless *other* compounds beyond vitamins that might be beneficial for your health, and which supplements and artificially fortified foods don't contain.

But there are definitely interesting questions to be asked about the broader implications of our reliance on synthetic vitamins. And if we were to ask these questions, we might find ourselves surprised by how political vitamins can become.

---

[2] There are several exceptions, including folic acid, which is slightly different from naturally occurring folate. (Synthetic folic acid is actually easier for our bodies to absorb.)

In the spring of 2001, the US Army found itself in the midst of a controversy that highlighted, albeit indirectly, an under-acknowledged fact about America's vitamin supply. The army's 225th anniversary was approaching, and the army chief of staff had decided that in celebration, every soldier in the army would be issued a black beret. When the army leadership went to order the berets, however, it ran into a problem: the Department of Defense was subject to something called the Berry Amendment, a piece of wartime legislation introduced in 1941—and repeatedly renewed—that required the DOD to buy food, fabric, and clothing (among other things) solely from domestic sources. But there weren't enough domestic beret manufacturers to produce the 4.8 million berets that this headgear shift would require. So the DOD got a waiver to outsource the making of the new berets to foreign countries, including China.

Then came an unfortunately timed collision between an American surveillance plane and a Chinese fighter jet. Tensions between the two countries were high, and—in one of those head-shaking political moments—the army's berets became a political cause célèbre. Eventually, the brouhaha became so intense that Deputy Defense Secretary Paul Wolfowitz was forced to issue an official announcement—the aptly titled "Deputy Secretary Wolfowitz Statement on Berets"—that commanded the army to cancel outstanding orders for Chinese-made berets, and to dispose of the ones that had already been delivered. The beret contracts were reissued to domestic suppliers (many of whom, ironically, got their source materials from abroad), and in late 2001, largely as a result of the beret controversy, the Berry Amendment was permanently enacted into law.

While critics complain that the Berry Amendment is anticompetitive and discourages free trade, its defenders claim that it's necessary for national defense—for if the United States were to be dependent on foreign sources for military supplies (or, I suppose, berets), it could be left vulnerable if those supplies were cut off. "[The] Berry Amendment serves as some protection for critical industries by keeping them healthy and viable in times of peace and war," explains one summary of the logic behind the legislation.

That brings us back to vitamins: the rules of the Berry Amendment also apply to military rations, products that are responsible for the health and viability of America's military members themselves. Most rations fulfill the military's nutritional requirements with the help of synthetic vitamins, making American military members dependent upon them for their essential nutritional needs. So where do those synthetic vitamins—as well as all the synthetic vitamins we civilians consume in foods and pills each day—come from?

I'll give you a hint: it's not America.

Back in the early days of synthetic vitamins, most breakthroughs in nutritional research took place in the United States or Europe. But while America played a crucial role in vitamin discoveries—and while plenty of *finished* foods and supplements are manufactured here today—it's never dominated the production of the bulk vitamins that are used as raw ingredients for these products. Instead, the Swiss health-care company Hoffman-La Roche maintained its status as the world's number one vitamin producer from the 1930s, when the first bulk synthetic vitamins were produced, all the way to the 1990s. By 1990, Roche still controlled about half of the global vitamin market; its nearest competitors were the German chemical giant BASF (which got its start in the 1860s producing dyes from coal tar), Rhône-Poulenc in France, and Japan's Takeda Chemical Industries. These four companies produced nearly 80 percent of the world's vitamins.

That level of industry concentration in substances that are fungible commodities is a recipe for corruption and collusion, and vitamin makers did not resist the temptation. In 1999, a dramatic press release from the US Department of Justice announced that Hoffmann-La Roche had agreed to plead guilty and pay a fine of $500 million for its role in organizing and running a global cartel that raised, fixed, and maintained prices for vitamins and vitamin premixes and divvied up market share for vitamin products. The scheme, which had started with vitamins A and E (and was directed by senior management), had expanded to include price fixing for vitamins B1, B2, B5, B6, C, D3, biotin, and folic acid—plus some additional carotenoids for good measure.

"This conspiracy has affected more than five *billion* dollars of commerce in products found in every American household," the assistant attorney general in charge of the department's Antitrust Division was quoted as saying (emphasis mine). "During the life of the conspiracy, virtually every American consumer paid artificially inflated prices for vitamins and vitamin-enriched foods in order to feed the greed of these defendants and their co-conspirators who reaped hundreds of millions of dollars in additional revenues."

Attorney General Janet Reno announced that Roche's $500 million fine in particular was "not only a record fine in an antitrust case, but it is the largest fine the Justice Department has ever obtained in any criminal case." BASF also pleaded guilty and agreed to pay $225 million for its role in the same conspiracy; Takeda paid $72 million, and many other smaller producers were involved as well. In 2001, the European Commission fined eight companies, including the ones mentioned above, almost a billion euros. All told, the *Wall Street Journal* estimates that the price-fixing lawsuits resulted in the bulk vitamin makers agreeing to pay $1 billion in criminal fines and more than $1 billion in civil judgments. According to Purdue economics professor John M. Connor's summary of the price-fixing scandal, "By the end of 2005, the members of these cartels had in absolute dollar terms become the most harshly punished antitrust violators in the history of the world."

There hasn't been any American-led major producer of synthetic vitamins since the 1980s, and today, in large part because of reorganization caused by this price-fixing bust, global vitamin manufacturing has shifted even farther away from the States. In 2002, Roche sold its vitamin business to the Dutch company DSM, which has relocated or closed all of Roche's US-based vitamin-manufacturing plants except for a beta-carotene facility in Freeport, Texas. BASF bought out Takeda's vitamin business in 2006, and as of 2013 had one US-based vitamin manufacturing plant: a vitamin E facility in Kankakee, Illinois. Rhône-Poulenc is now part of Sanofi (formerly Aventis), which doesn't manufacture any vitamins in America. Daiichi Fine Chemical, another large Japanese producer, was bought in 2007 by the Japanese pharmaceutical company Kyowa Hakko Kirin, which makes vitamin K at a plant in Japan (but nothing in the United States). Other Western producers like Eastman Chemical, Degussa, Merck, and Eisai left the vitamin business entirely, and the remaining players closed most of their large production facilities for vitamins E and C in the United States, shifting them to Europe or Asia instead.

The global vitamin market is currently dominated by two European companies, DSM and BASF, but their main competitors—and indeed, many of their own production facilities—are now in China.[3] Most of the world's supply of vitamins A, B12, and E comes from China, along with about 75 percent of vitamin D and more than 80 percent of vitamin C. According to a 2011 report by Leatherhead Food Research, China exports between 150,000 and 200,000 tons of vitamins per year, up from fewer than 100,000 tons in 2003.

China, where several companies account for 70–90 percent of the country's market share, is no more immune to price-fixing than its European counterparts. In the mid-1990s, thanks to a new production method, Chinese vitamin C flooded the market, driving prices down and helping to break some of the European cartels—and then in March 2013, a jury in New York found a group of Chinese vitamin C makers guilty of price-fixing and ordered them to pay a total of $162.3 million in fines. The accused companies claimed that the Chinese government had ordered them to fix prices, but the jury was unconvinced. According to the plaintiffs' lawyers, the Chinese vitamin C makers found guilty of price-fixing were already functioning by 2001, the same year that the European Commission fined the participants in the original vitamin cartels. Far from being discouraged by the American and EU cases, the Chinese cartels had flourished.

These statistics (not to mention the price-fixing cases themselves) are largely unknown to consumers. Indeed, when an industry trade group called the United Natural Products Alliance surveyed a thousand American supplement consumers in 2007 and again in 2011, it found that the average American consumer believed that less than 10 percent of the world's supplement ingredients came from China. In addition, more than 63 percent of Americans said that if a supplement's ingredients *did* come from China, they'd be less likely to buy it.

It's important to note that quality concerns can arise anywhere, and our

---

[3] Now that the majority of Europe's, America's, Australia's, and New Zealand's wool-washing factories have closed, most greasy wool—the raw material from which lanolin is derived—ends up in China. There, according to a brief article in one of my new favourite publications, *Anhydrous Lanolin News*, "wool can be washed at lower costs and a strongly growing local market for woolen textiles is to be served." As a result, whereas in 1999 less than a quarter of Australia's wool was washed in China, by 2010 analysts expected the country to have an 80 percent share in "Australian greasy wool processing," making China the wool-processing—and vitamin D-producing—capital of the world.

bodies don't care which country their vitamins are from. But, getting back to the Berry Amendment, the lack of domestic vitamin producers does raise an interesting political point. The amendment requires the military to source its rations from domestic manufacturers. Producing synthetic vitamins, unlike berets, requires sophisticated and expensive production facilities that have been designed specifically for each vitamin; it's not easy to open a new factory or restart one that's been closed. And again, unlike berets, vitamins are essential for human—and, therefore, service members'—health. If you're going to argue that America is putting itself at military risk by not having domestic beret-making facilities, then shouldn't you also be concerned about where it's getting the micronutrients for those beret-clad service members' food?

As it turns out, the companies that produce America's rations get around the Berry Amendment with the help of a loophole: ingredients can be sourced from abroad as long as the end products are manufactured in the States. Given the fact that there are nearly no domestic manufacturers of bulk synthetic vitamins, it's a critical exception. Today, if the army decided to celebrate its anniversary by issuing special made-in-America multivitamin packs to all its soldiers, it would be impossible for American businesses to fill the order.

If you continue with this potentially xenophobic thought experiment, then you'll soon encounter an even bigger question: If the military is dependent on foreign companies for vitamins, then how about the general public?

When the American public first became aware of vitamins in the 1910s and 1920s, the country's food supply was largely capable of providing sufficient quantities of natural vitamins from unfortified food: the prepackaged and fast-food businesses had not yet exploded, the population was smaller, and people cooked many of their meals themselves, using ingredients whose vitamins and other micronutrients had not been destroyed by processing and refinement. As Dr. M. L. Wilson of the US Department of Agriculture put it to the *New York Times* in 1941, "We have the productive power in our agriculture to supply the adequate diets needed for all." The same article pointed out that the government was toying with the idea of improving the nutritional quality of flour not via synthetic vitamins, but via improved milling methods that ground up a higher percentage of the wheat grain (85 percent compared with the 60 70 percent typically used in highly refined flour), which would have preserved more of the wheat's natural vitamins. As its author explained, "The government experts don't care how it is done as long as it is done."

It's possible to imagine a historical trajectory where, once vitamins' importance had been accepted, government policy and market demand would have created a food supply very different from what we have today. In this alternate reality, vitamin-dense foods like fresh produce would be subsidized rather than soybeans, wheat, and corn, which are the cornerstones of modern processed foods and whose natural vitamins—as we'll see—are mostly removed or destroyed before we eat them. Could the nutritional needs of the rapidly growing American population have continued to be met without the use of synthetic vitamins? Perhaps not, especially given how many of us there now are. But there's no way to know—because that's not the direction we decided to take.

Instead, twentieth-century food scientists focused on developing ways to ensure longer shelf lives. Chemists came up with more than four hundred new additives to help in the processing and preservation of food from 1949 to 1959 alone, and by 1953 packaged, processed foods had become so popular that *Fortune* magazine noted that Americans' "relentless pursuit of convenience" meant that "there are few jokes these days about young brides whose talents are limited to a knowledge of the can opener . . . 16 billion pounds of canned goods are now going down the national gullet every year." Agricultural scientists were also more focused on price and ease of processing than on enhancing the nutrient content of animals and crops. Challenges like the 1945 Chicken-of-Tomorrow contest, for example, were designed to create a bird that was cheaper to produce and easier to market, not one that was more nutritious. If any of these advancements came at the cost of decreasing the foods' vitamins—and if these foods' creators cared about the issue to begin with—technology provided an easy solution: synthetic vitamins were added back to replace those that had gone missing. Unfortunately, this quick fix ignored the fact that food contains countless other chemicals besides vitamins that may be beneficial to our health, which processing can destroy—and which are usually not replaced.

Presumably food companies' and scientists' priorities were driven by consumer demand—as the head of food science at the University of California,

Davis, put it, "If food isn't safe, convenient, good to eat and resistant to spoil-age, most people would throw it out regardless of its nutritive value." Neverthe-less, the result is that today not only do most of our synthetic vitamins come from abroad, but much of our vitamin-rich produce does as well: more than half of America's fresh fruit comes from overseas, as do many of our vegetables. If we truly wanted to be nutritionally independent, whether as a military or a country, we would have to change the way our agricultural system works.

Of course, many of us don't seem to gravitate toward produce or other foods that are inherently high in vitamins no matter which countries those foods come from. Instead, faced with the choice of eating foods that are nutrient-rich or eating the foods we crave, we demand a third option: to do both. We want to eat our cake (or breakfast cereal or toaster pastry) and get our vitamins, too. And thanks to the abundance of cheap, synthetic vitamins available from abroad, we're able to do just that.

The primary reason that food companies are willing to provide us with so many vitamin-enhanced products is simple: they're profitable. Synthetic vita-mins produce a tremendous bang for the food manufacturers' buck, essen-tially conjuring value out of thin air. As I write, commodity wheat costs about 12 cents per pound, a pound of sugar can be had for 42 cents, and I can per-sonally buy a kilogram of vitamin C off Amazon.com for twenty-four dollars (an amount that's equivalent to the RDA for more than eleven thousand adult men). In contrast, an eighteen-ounce box of Total cereal, which is basically wheat with sugar and vitamins, is $5.59 from FreshDirect.

But there's another reason for the prevalence of vitamin-enriched and vitamin-fortified processed foods that's rarely, if ever, discussed: these prod-ucts exist because we *need* them to. In the aforementioned study from the *Journal of Nutrition* ("Foods, Fortificants, and Supplements: Where Do Amer-icans Get Their Nutrients?"), the authors concluded that "[w]ithout enrich-ment and/or fortification and supplementation, many Americans did not achieve the recommended micronutrient intake levels set forth in the [official dietary recommendations]." Translation: without synthetic vitamins, we'd be in trouble.

The fact that we *don't* generally suffer from severe nutritional deficiencies is due to a subtle form of intervention: although no micronutrient enrichment or fortification is currently mandatory in the United States, several products have been vitamin-enhanced for so long that we don't even recognize them as such.[4] Bread is often made from flour that's been voluntarily enriched and fortified with thiamin, niacin, riboflavin, and iron. Most milk has been fortified with D for such a long time (beginning in 1933, and originally accomplished by irradiating the milk with ultraviolet light or feeding cows irradiated yeast) that it's become a major dietary source of vitamin D without most of us realizing that it's an artificial addition. And breakfast cereals? Let's put it this way: Cap'n Crunch's Crunch Berries are not getting their nutritional boost from fruit.

In cases like milk and cereal, where fortification and enrichment have occurred for so long that they've become invisible, it would be strange—perhaps even irresponsible, from a public health perspective—*not* to fortify them. If food companies didn't voluntarily do so, the government might have to require it, to make sure that we don't accidentally eat ourselves into nutritional deficiency.

But this won't happen, because synthetic vitamins are as essential to food companies as they are to us. If processed products were not enriched with synthetic vitamins, the nutritional emptiness of their raw ingredients would mean that we'd have to eat (or take) something else to meet our micronutrient requirements. In fact, it's possible that without synthetic vitamins, the selection of packaged foods in the middle aisles of today's grocery stores would never have been able to grow so large—and the modern grocery store as we know it would not even exist.

Either way, the result is that we've created an odd symbiotic relationship, in which companies depend on us to buy their products, and we depend on the synthetic vitamins in these products to fulfill our nutritional needs. This keeps us from becoming deficient, but there's a consequence: not only are we missing out on whatever other important dietary chemicals might be present in unprocessed food, but the constant supply of synthetic vitamins blinds us to our own dependence on them.

---

[4] If there's widespread risk of deficiency in a particular nutrient, then targeted fortification and enrichment can be powerful tools to prevent that deficiency. As a mineral example, the addition of iodine to salt, which began in 1924, has enormously reduced the rate of goiter and mental retardation caused by iodine deficiencies.

And we are indeed dependent. We like to believe our food system is the most sophisticated in the world, but statistics like those from the study mentioned above suggest that our pride is hollow; without synthetic vitamins and the products that contain them, we'd be as susceptible to deficiency diseases as the societies to whom we provide nutritional aid. If it weren't for our easy access to man-made sources, the conversation about nutrition—not to mention our food supply—would likely be quite different.

Indeed, the so-called standard American diet—high in refined grains and sweets, and associated with "Western" diseases like heart disease and cancer—could not have developed without the help of synthetic vitamins. This has led to an odd paradox. Given the limitations of the global food supply and consumer preferences, synthetic vitamins are truly essential for the prevention of nutritional deficiency diseases—not just in the developing world, but here, too. In the West, however, where synthetic vitamins are widely used to correct nutritional deficits caused by processing, they've also contributed to the very problem they were meant to fix. While they're designed—and now often required—to keep us healthy, synthetic vitamins also enable the very products and dietary habits that are making us sick.

That itself is unnerving. But vitamins haven't just shaped our food supply; they've also shaped our minds. More so than any other component of food, vitamins are responsible for our current approach to nutrition, a perfectionist attitude that's simultaneously misguided and fantastically naïve. As such, it's worth exploring how vitamins were discovered and how the public came to embrace them—in hopes that if we understand the history of our philosophy toward nutrition, we might be able to improve it going forward. This story of scientific progress didn't begin in a lab but in the lush landscape of nineteenth-century South and Southeast Asia, where people had begun dying from a mysterious and terrible disease.

3

Death by Deficiency

Finding a needle in the proverbial haystack is far easier than
cornering and then isolating a vitamin. The needle-hunter
knows at least what haystack he must pull apart, but the
vitamin-hunter must even find the haystack.
—WALDEMAR KAEMPFFERT, "WHAT WE KNOW ABOUT
VITAMINS," *New York Times Magazine*, 1942

n 1814, a British army surgeon named J. Ridley travelled to a small jungle
garrison in the colony of Ceylon, now Sri Lanka, to take care of native
troops who were suffering from a strange sickness. Known as beriberi, it was
frequently seen in South and Southeast Asia, but didn't normally affect foreign officers—something appeared to grant them immunity. For the natives,
however, it was deadly. It began with intense swelling of the legs and feet and
a general sense of numbness, especially in the extremities. Victims developed
a distinctive gait, lifting their knees high in the air and swinging their legs
forward so that their drooping toes wouldn't catch on the ground. Their urine
became concentrated and brightly coloured, and they lost their appetites even
as their bodies wasted away. As the swelling increased, victims began to feel
a pressure under their ribcage so intense that they sometimes "solicit[ed] that
the part may be cut open," wrote Ridley, "expecting to have the tightness
relieved by that means." Eventually, they lost their voices and died in suffo-
cating convulsions.

Beriberi's cause was a mystery. It came—and still comes—in two primary,
sometimes overlapping forms, with the nervous system symptoms referred to
as "dry" beriberi, and the cardiovascular damage as "wet" beriberi. Many of
Ridley's patients believed it to be the work of a devil, but Ridley thought that
the disease was more likely due to some combination of bad water, something
toxic in victims' diets, and damp air.

We now know that neither a devil nor dampness is to blame. Instead, beriberi's dramatic symptoms are the result of a deficiency in thiamin, a bitter

vitamin also known as B1 that's found in foods including yeast, grains, nuts, and meat. These days, beriberi has become so uncommon in the developed world that few of us even recognize its name. But beriberi played a crucial role in kick-starting the process of scientific inquiry that led to vitamins' discovery. It was through the study of beriberi that the idea of nutritional deficiency diseases—a concept essential for the discovery of vitamins—first began to crystallize. And it was the anxiety provoked by the possibility of these nutritional deficiencies that has come to define how we think about nutrition today, an obsessive relationship well summarized by the Hippocratic saying "Let food be thy medicine, and medicine be thy food."

In Ridley's time, no one had ever heard of a nutritional deficiency disease, let alone a vitamin; even the discovery of disease-causing germs (and, therefore, bacteria-borne diseases) was still decades away. And so Ridley struggled to treat patients on a case-to-case basis with no effective remedy, in conditions that were quickly spiralling out of control.

Men were dying at a rate of five to eight a day; the sick were lying in filthy cots, surrounded by fly-covered piles of faeces and vomit. Clean water was scarce, and the garrison's wells had to be guarded from a hearty population of wild elephants, who were "attracted thither from the neighbouring jungle, where they were in immense numbers." Since the only other European had died of jungle fever just a few days after Ridley's arrival—and he didn't trust the natives, whom he accused of having a "natural laziness"—Ridley supervised all the work himself. He had the wards scrubbed and fumigated, gave doses of laxatives and diuretics, and ordered the patients' legs and feet to be bathed in warm water, rubbed, and wrapped. In an effort to catch the disease in its early stages, he examined the healthy men twice a day for any signs of it.

Convinced of his own immunity as a foreigner, Ridley hopped from patient to patient, trying to deal with a rate of death so high that "it occurred, more than once, that some of those who attended the funeral of their comrades one evening, were themselves followed to the grave the next." He worked around the clock for nearly two weeks, grabbing food and sleep when he could. Ridley was so busy that at first he tried to ignore it when he noticed some strange symptoms in himself—fatigue, trouble breathing, and a sense of heaviness in his limbs. But on the morning of the thirteenth day, he was forced to admit that something was seriously wrong.

"I awoke with a sensation of tightness, as if a bar were placed across my breast, and impeded the action of my lungs," he wrote. "Upon getting up . . . I found my legs and feet perfectly numbed, swoln [sic] and edematous; my lips were numbed, and felt unusually enlarged; and the space round my mouth, reaching nearly to my eyes, felt numb."

Ridley had beriberi.

He took a dose of an opium tincture, plus some brandy and a laxative, but they didn't help—his symptoms grew worse and his face and throat began to swell. Terrified, he had his servants carry him on a litter back to his base, nearly a hundred miles away; they frequently had to stop and help him sit upright to overcome fits of breathlessness. He made it to a doctor, who enabled him, via an unspecified treatment, to partially recover. But soon afterward, the disease returned again—this time causing vomiting, and an "extraordinary fluttering of the heart" that he experienced whether he was reading, walking, or sitting "perfectly still." (He claimed it was possible to see the pulsing of his carotid arteries from five yards away.) He eventually recovered after being transferred to another garrison, but was left so sick and weak that he was forced to return to England. Later, Ridley reported that the first attack had left his memory considerably impaired, and that even five years afterward it "ha[d] not been completely restored."

Scientific efforts to understand beriberi's cause and cure began in earnest later in the nineteenth century, some sixty years after Ridley had developed the disease—by which point beriberi's incidence, particularly in Asia, had exploded. These researchers weren't studying vitamins, however, for no one knew that such a thing existed. Indeed, they weren't even studying human nutrition, let alone looking for some food-related magic bullet. Instead, inspired by the most exciting medical event of the century—the discovery of disease-causing germs—these scientists were on the hunt for a beriberi-causing bug.

They were hardly alone. Thanks in large part to the enthusiasm for this germ theory of disease, it would take decades before scientists recognized and accepted the essential premise of all nutritional deficiency diseases: that sickness can be caused not just by the presence of something bad, but by the absence of something good. And it would take longer still to discover that this "something good" was a group of invisible compounds found in food.

When Westerners began to visit Japan in the mid-nineteenth century, it must have been a culture shock on both sides—Japan had largely cut itself off from the Western world for the previous two hundred years, with severe restrictions on allowing foreigners to enter or its own citizens to leave. But among the myriad new sights that surrounded them, the first Western physicians to arrive in Japan encountered something unexpectedly familiar: an illness known in Japanese as *kak'ke*—"leg disease"—that looked remarkably like the beriberi they'd seen elsewhere in Southeast Asia.

There were many guesses about what caused the affliction, including the hypothesis that beriberi was caused by poisonous air rising up from wet soil (a version of the miasma theory of disease, which held that disease was caused and spread by poisonous, foul-smelling airs). This made some sense: beriberi was usually limited to particular geographic areas and didn't seem to spread from person to person, and the paralysis often started from the feet—the body part closest to the earth. It was certainly just as plausible as some of the other theories Western doctors proposed, including that beriberi might be caused by sexual excess or the Japanese habit of squatting on the floor instead of sitting in chairs.

Around the same time that it opened its borders, Japan realized that if it wanted to remain independent from the European powers trawling the Pacific, it needed a strong navy. Great Britain's navy was considered the best in the world, so Japanese naval officials ordered ships from Britain, appointed a British surgeon as a professor of medical science in the naval college, and set up a private medical school in Japan taught by British instructors. One of these naval students was Kanehiro Takaki, a young surgeon recruited into the navy around 1870 who was also sent abroad to do five years of postgraduate work in London under the guidance of a British doctor. Upon his return to Japan in 1880, Takaki began to study the problem of beriberi in the Japanese navy. It was a huge concern: from 1878 to 1882, roughly a third of enlisted men reported becoming sick with it—many fatally so—each year.

The "poisonous soil" idea didn't grab Takaki, likely because of the lack of any soil on ships, poisonous or otherwise. He also noticed that European sailors living in similar conditions didn't often suffer from beriberi. In 1883, Takaki inspected ships and naval barracks and concluded that while factors like working hours, cleanliness, and clothing were relatively consistent, there were considerable differences between the diets of the Europeans and the beriberi-prone Japanese. This led him to suspect a nutritional connection. At the time, food was thought to consist of water and the three macronutrients: protein, fat, and carbohydrate (minerals were acknowledged but no one knew their nutritional significance). And so, unaware of vitamins, Takaki came up with a hypothesis that fit with the nutritional understanding of the day: that beriberi was related to protein consumption, since the white-rice-loving Japanese sailors were eating far less of it than their beriberi-free English and German counterparts.

A particularly bad cadet-training voyage to New Zealand provided the impetus to test out his theory. Out of the 278-person crew, more than half the men had become sick with beriberi and 25 had died. Takaki persuaded his superiors to repeat the trip but with modified rations: instead of rice, he proposed a protein-heavy diet of meat, condensed milk, bread, and vegetables. Although beriberi is not actually related to protein consumption, the experiment was still a success—the diet happened to contain enough thiamin to keep beriberi at bay. No one died, and the only people who showed signs of the disease later admitted they hadn't eaten all the new rations.

Takaki persuaded the navy to alter its sailors' diets, and in 1887 he reported that not a single sailor had died of beriberi that year, compared with more than a thousand a year before the change took place. He was rewarded with a personal interview with the Japanese emperor and, later, the rank of baron. Though his protein hypothesis was incorrect, Takaki had solved the Japanese navy's beriberi problem, an achievement for which he's still celebrated today.

Despite Takaki's success with the navy, however, his dietary changes were not adopted by the army, which continued to suffer from beriberi for years to come. Part of the army's hesitancy to accept Takaki's dietary changes likely came from the fact that while the Japanese navy had reached out to the British for training, the army had instead recruited German scholars. This was a crucial difference. First, in the late 1800s, the very decades in which the incidence of beriberi began to explode, Germany was considered the world's leader in nutritional research. Its scientists were convinced that all of the important chemical components of food had been figured out: fat, carbohydrate, protein, water, and what we now know as minerals—leaving no conceptual room for the idea of vitamins. And second, Germany was home to some of the world's leading researchers in germ theory, who were discovering specific microscopic organisms—known as pathogens—that were responsible for many of the world's most terrifying illnesses.

Before the spread of germ theory, people recognized that certain diseases were connected to certain circumstances—cholera seemed somehow linked to water, for example. But no one knew what, exactly, was behind these connections. Various scientists had previously proposed that sickness might be caused by little creatures too small for the eye to see, but besides a few fungal infections, no one had proved a definitive link between a microorganism and a disease. The ancient theory of the humours still lingered, and many diseases were still attributed to miasmas. The concept of spontaneous generation—that living organisms could spontaneously appear out of nowhere—only added to the confusion. No one could tell whether microorganisms gathered from sick people were the cause of an infection or the result.

In 1862, the French chemist and microbiologist Louis Pasteur famously disproved the idea of spontaneous generation by demonstrating that meat broth would not grow cloudy with microorganisms if it wasn't exposed to air. He also showed that spoiled milk was caused by the proliferation of bacteria, and in 1863 he invented the sterilization process that still bears his name. Inspired by the idea that eliminating microbes could prevent infection, the British surgeon Joseph Lister established many of the sanitary medical practices that we still follow today, like prepping with alcohol swabs and hand washing. (His legacy also includes doctors' white coats—which promote cleanliness by revealing dirt—the white bathroom tiles that were popular at the turn of the twentieth century, my aversion to touching subway poles, and the inspiration for the name Listerine.) In 1876, the German physician Robert Koch proved that anthrax was caused by a bacillus; in 1882, he discovered the bacteria that caused tuberculosis and, a year later, cholera.

While it took a long time for germ theory to be universally accepted, over the next twenty-five years—precisely the time that investigations into beriberi (and, as we'll see, nutrition) were getting off the ground—scientists identified the microorganisms responsible for diseases including diphtheria, typhoid, tetanus, syphilis, gonorrhea, pneumonia, and bubonic plague. Malaria turned out to be caused by a parasite. In the 1890s, the concept of viruses (extremely small infective molecules that can only reproduce within the cells of a host) was established. It was like a gust of wind that sweeps aside a cloud: one idea—that tiny organisms could cause infection—revealed an entire sky's worth of possible explanations.

Pro-germ theory scientists were understandably giddy. For the first time in history, it was possible, with the help of a microscope, to actually *see* the causes of many of the world's most terrifying diseases. The eventual acceptance of germ theory was inarguably one of the greatest medical advances in history, leading to the prevention of many diseases and encouraging scientific curiosity about the cause of sickness instead of just the cure. For nutritional science, however, its impact was more complicated. Whereas people had previously tried to find a humoural explanation for all disease, they now did the same with germs. In many cases, they were successful. But germ theory's central tenet—that disease is caused by the *presence* of something—hid the idea that disease could also be caused by something that is *lacking*. Germ theory's light was so bright, so illuminating, that it blinded scientists to the idea that disease could be caused by something that wasn't there.

When the Dutch physician Christiaan Eijkman arrived on the Indonesian island of Java in 1886 (by then a Dutch colony) to investigate the causes of beriberi, he didn't know that he was about to launch the line of research that eventually identified beriberi as a vitamin deficiency disease—let alone that his efforts would win him the first vitamin-related Nobel Prize. Nor did he know about Takaki and the contemporaneous work going on within the Japanese navy. Instead, true to the spirit of the day, Eijkman's assignment was to find the pathogen that caused beriberi. It was a task for which he was well prepared. Not only was he already familiar with beriberi—he had been living in Indonesia before malaria forced him to return to Europe—but he had recently trained in the new field of bacteriology in Berlin at the laboratory of Robert Koch.

Working in a civilian research unit in an army hospital in what's now Jakarta, Eijkman decided to use chickens as his laboratory animals. He chose them mostly because they're easy to raise in large numbers, but the decision was lucky, since chickens and pigeons are two of the only animals other than humans that frequently develop the disease (other common lab animals like dogs, rats, monkeys, and rabbits are less susceptible). He procured a flock and, once his experimental subjects had settled into their cages, he began injecting them with blood samples from human beriberi patients to see if he could infect the birds.

After a couple of months, he saw symptoms in some of his injected chickens that looked a bit like the nerve damage that occurred in people with beriberi. Then again, he saw the same symptoms in his control group. But Eijkman was not deterred—after all, many pathogen-borne diseases can be transmitted by air, and the two groups of chickens had shared cages. He got some new chickens and put them in their own individual bamboo cages. The controls still developed the beriberi-like nerve damage, known as polyneuritis (multiple inflamed nerves). Concluding that perhaps his whole laboratory had become infected, Eijkman procured yet another group of chickens and kept them in a totally separate location. Then things got really strange: not only did none of the new chickens develop polyneuritis, but the sick birds began to recover. By November 1889, all signs of the disease had disappeared.

This bizarre mass recovery might have been good for the chickens, but it was bad for Eijkman, who appeared to have lost his animal model. But whereas other researchers would have thrown up their hands or switched to a different species, Eijkman did not give up on his chickens. Instead, he tried to find any possible variable that could have accounted for the sudden change. One day, the laboratory keeper told him something intriguing: that in the month before the birds developed polyneuritis, the cook had been providing leftover white rice from the hospital's kitchen as their feed.

At the time, white rice, otherwise known as polished rice, was something of a luxury—or at least not something you'd give to laboratory chickens that you were simultaneously trying to infect with a deadly disease. (Animals usually ate brown rice.) That cook had been replaced, and his successor, Eijkman later related in his 1929 Nobel lecture, "refused to allow military rice to be taken for civilian chickens." So the birds had been switched back to their usual rations of brown, unpolished rice; soon thereafter, the polyneuritis disappeared.

The terms "polished" and "unpolished" refer to how the rice is milled. In its natural state, rice has a tough, indigestible husk that you need to remove before you eat it. Take off the husk, and you're left with brown rice, whose colour comes from a second interior skin called the pericarp, also known as the polishing. Take off the pericarp, and you're down to the endosperm—the white, polished rice kernels that we're familiar with today. Low in fibre and mostly starch, the endosperm's purpose in the plant is to provide the energy necessary for the rice seedling to grow.

In the days before milling machinery, people milled rice by hand, which made the rice digestible but left pieces of the pericarp behind. Beriberi occasionally occurred, but was relatively rare. Then, around 1870, European colonialists brought mechanized steel rollers to Asia, machines that were both faster and removed the polishings more efficiently. Mechanized milling produced the desired white rice, but the colour came at a cost: rates of beriberi began to skyrocket. Diets that relied too heavily on polished rice seemed to make people—not to mention Eijkman's chickens—sick.

But even though he noticed that the timing of the onset and remission of his chickens' polyneuritis precisely matched the change in their diets, Eijkman was hesitant to accept the idea of a dietary connection. This was partially because he was so wedded to finding a bacterial explanation, and partially because he wasn't yet sure if the disease he was observing in his chickens was the same disease as human beriberi. He called it polyneuritis gallinarum to be safe (*gallus* is Latin for "rooster").

The nerve damage caused by the two diseases was suggestively similar, however. In polyneuritis gallinarum, the bird's gait becomes unsteady and it's unable to perch. Its legs become so weak that they spread apart, making it look like it's attempting a split. Soon it can't walk. As paralysis creeps upward, the bird can no longer move its head; its breathing slows and its beak opens. Photographs of polyneuritic pigeons (which later experimenters used instead of chickens) show a pathetic sight: birds whose necks are bent so far backward that their throats form an upside-down U and their beaks face the ceiling, as if they're in some horrible contortionist act—a condition known, very euphemistically, as "star-gazing." Leave them untreated, and they'll die.

Today, we know why the polished rice caused problems. Rice polishings—and, indeed, the outer coatings of many whole grains—contain thiamin, among other vitamins and nutrients. The better the milling process, the lower the level of thiamin that remains. Additionally, we now know that while beriberi is often closely linked to diets heavy in white rice—indeed, the disease still occurs in South Asia, as well as in prisons and other confined situations where people eat thiamin-poor diets—a thiamin deficiency can occur in many other circumstances as well. Thiamin deficiencies are particularly common in alcoholics, and can also be exacerbated by diets high in refined carbohydrates that haven't been enriched to replace their micronutrients, since thiamin requirements increase in line with carbohydrate consumption.

Thiamin plays a crucial role in breaking down carbohydrates, synthesizing RNA and DNA, and maintaining the brain and nervous systems. Many of the precise details of how a deficiency in thiamin actually *causes* beriberi's symptoms are not fully understood, but the progression of various forms of the disease has been mapped out. For example, in wet beriberi—the version that primarily affects the cardiovascular system and enlarges the heart—a lack of thiamin causes peripheral blood vessels to dilate, and the resulting lower blood pressure makes the heart work harder. (Think of how much more energy it takes to pump water through a fire hose compared to a garden hose.) The kidneys, meanwhile, erroneously interpret this lowered blood pressure to mean that there's a low volume of blood in the body overall, so they begin to retain salt. The salt pulls fluid into the blood, increasing its volume. This extra fluid then causes wet beriberi's characteristic symptoms of swelling, particularly in the hands and feet. It also makes the heart, which by now is likely enlarged and at risk of injury from overuse, work even harder.

It's possible for a thiamin deficiency to cause irreversible damage, including death. But if caught early enough, the effects of treatment can be dramatic. In its pure form, a dose of one-hundredth of a milligram of thiamin a day is enough to cure a deficient pigeon. Give thiamin to a person with wet beriberi and he can begin to show improvement in hours; his heart will be back to normal in one to two days. Yet despite its importance—and probably because so many foods naturally contain it—human adults only store about 25–30 micrograms of thiamin in their bodies. Since it has a half-life of between ten and twenty days, thiamin depletion can occur within weeks.

Eijkman did eventually accept that there was some sort of connection between white, polished rice and the chickens' disease, but he still thought it must be related to bacteria. After learning that some forms of polyneuritis in humans are caused by poisoning from bacteria-produced toxins, he concluded that there must be a beriberi bacterium in the white part of rice that was producing a poison—and that there was some sort of "anti-beriberi factor" in the rice polishings that was an antidote to these toxins. Removing the polishings meant losing the antidote, argued Eijkman; that's why white rice caused beriberi.

Regardless of the bacteria issue, in 1895, shortly before malaria would force him back to the Netherlands for good, Eijkman finally got a chance to answer the question of whether human beriberi and polyneuritis gallinarum were the same disease. This chance came by luck when he struck up a conversation about his beriberi/polyneuritis gallinarum investigations with a friend of his who was the medical director of all the prisons in Java (the island had 101 prisons and about 250,000 prisoners). Said friend realized that different prisons on the island fed inmates different types of rice, and that prisons varied in how many cases of beriberi they reported. This data could be used to determine whether rice had anything to do with human beriberi— and, therefore, whether Eijkman's chicken work was relevant to the human disease.

Preliminary results indicated that beriberi and rice were indeed connected, and when his friend conducted a more intense follow-up analysis after Eijkman's departure, he found that while only 1 out of 10,000 prisoners developed beriberi in the prisons that served mostly brown rice, 1 out of 39 developed it in those that served white. Among long-term white-rice-eating prisoners, the rate went up to 1 out of 4.

This might seem like convincing evidence of a dietary connection, but when Eijkman suggested it to his colleagues back in the Netherlands, they mocked him. "If one considers that Eijkman apparently needed six years in order to do this work, it must be considered the most inadequate product which can be found in the literature from the Director of a scientific institute," wrote one charming colleague. When Eijkman proposed to a different colleague that perhaps pontificating on the cause of a disease should be left to those who had actually studied it firsthand, the man responded that Eijkman had likely suffered brain damage from eating too much rice. Little did that critic know that, while Eijkman did not personally figure out that beriberi was a deficiency disease, he would receive the 1929 Nobel Prize in Physiology or

Medicine for his recognition of this "anti-beriberi factor," which by then was known as thiamin, and for developing research methods that influenced later nutritional scientists and advanced vitamins' discovery.

While no single person can truly be credited as "discovering" thiamin, it was Eijkman's successor, a fellow Dutch researcher named Gerrit Grijns, who wrote what's often considered to be the first correct hypothesis on beriberi's cause—suggesting the existence of not just what we now call nutritional and vitamin deficiency diseases, but also specific substances in foods that could prevent them. (Grijns never worked directly with Eijkman and was not wedded to the idea that beriberi was caused by microbes.) As Grijns famously wrote in a paper published in 1901, after four years spent methodically researching the condition, "There occur in natural foods, substances which cannot be absent without serious injury to the peripheral nervous system. The distribution of these substances in different foodstuffs is very unequal."

Grijns had correctly identified beriberi as a nutritional deficiency disease, caused by the lack of a chemical substance only present in certain foods. Unfortunately, he wrote his hypothesis in Dutch and it wasn't translated into English or published internationally. As a result, this statement didn't become widely known for about twenty-five years, well after the concept of vitamins had been accepted. Likewise, another Dutch researcher named Cornelius Pekelharing published a similar observation in 1905, claiming that "[t]here is a still unknown substance in milk" that was essential for life, but it was two decades before this work was translated into English.

From our vantage point, it's tempting to scoff at how long it took people to accept the idea of nutritional deficiency diseases. But as Gerald Combs, author of the textbook *The Vitamins*, pointed out to me, each of these researchers was working "with a fraction of the knowledge we have now and interpreting their findings in the light they had." As he put it, "[T]he process of elucidation of the vitamins—indeed, the process of scientific discovery—is a human endeavor fraught with fits and starts and intellectual cul-de-sac."

Perhaps stranger is that even after vitamins were suspected and the idea of nutritional deficiency diseases was widely acknowledged, germ theory *still* got in the way of their acceptance. No story better exemplifies the enduring

distraction caused by germ theory than that of pellagra, a disease that terror-
ized the American South less than eighty years ago. The rigid thinking that
blocked the discovery of its cause can be seen as a cautionary tale for us today.

Originally described by a Spaniard in 1735, the first case of pellagra in the
United States was recorded in 1864. It quickly became an epidemic, especially among poor southern sharecroppers and asylum patients—some of
whom may have landed in the asylum *because* of pellagra, since it affects the
brain. The disease, which had a mortality rate above 50 percent and by 1911
had become the leading cause of death in asylums, was so feared that physicians and nurses at Johns Hopkins in Baltimore were discouraged from even
saying the word.

Pellagra is caused by a deficiency in niacin (vitamin B3) and, like beriberi,
it's no longer a household term. But the fact that it's been forgotten belies the
devastation it caused and the drama of its cure. An Italian physician named
Francesco Frapoli coined the term "pellagra" (it comes from the Italian for
"rough skin") and wrote the following description in 1771:

> [T]he colour of their skin changes suddenly to red . . . frequently small
> tubercles of varied colour rise up; then the skin becomes dry, the sur-
> rounding coats burst, the affected skin falls in white scales just like bran;
> finally the hands, feet, chest, rarely even the face and other parts of the
> body exposed to the sun become repulsively disfigured . . . the disease
> rages recurrently until at length the skin no longer [peels off] but becomes
> wrinkled, thickened and full of fissures. Then for the first time the
> patients begin to have trouble in the head, fear, sadness, wakefulness and
> vertigo, mental stupor bordering on fatuity, hypochondria, fluxes from the
> bowels, and sometimes to suffer from mania, then the strength of the body
> fails, especially in the calves and thighs and they begin to lose motion of
> those parts almost entirely, to emaciate in the highest degree, to be seized
> with a . . . diarrhoea most resistant to all remedies and consumed with a
> ghastly wasting, they approach the last extremity.

In 1914—by which point the idea of deficiency diseases had gained traction and the word "vitamin" had been coined—an American doctor named

Joseph Goldberger was sent to the South by the US Public Health Service to discover the cause of pellagra. At the time, one of the leading theories was that it was an infectious disease transmitted by a "pellagra germ." (The other leading theory was that it was caused by a bacteria-produced toxin in corn that was especially prevalent when the corn got moldy.) The fact that its damage to the skin was somewhat similar to that of syphilis and leprosy, both found to be caused by bacteria, bolstered the germ hypothesis.

Goldberger disagreed. One of the occupational hazards of his career was that he spent lots of time in close contact with sick people, and he liked to think he knew a thing or two about the patterns of infectious disease. Pellagra didn't fit the pattern: people who cared for pellagrins, as patients were called, didn't come down with pellagra themselves. Goldberger proposed in a 1915 report that pellagra was somehow caused by diet—in particular, the southern staples of "the three Ms": meat (pork fatback), meal (cornmeal), and molasses. Unlike Eijkman, he embraced the idea that a disease could be caused by a nutritional deficiency and never tried to explain the dietary connection with a germ. But his theory was rejected, partially because of the popularity of germ theory and partially because of politics—he was a Jewish European immigrant, raised in the North, who had been sent to the South uninvited, and now appeared to be criticizing their way of life.

Undeterred, Goldberger spent the following decade designing rigorous studies that showed pellagra was more prevalent among people who ate corn-based diets, that it could be cured by eating different food, and that it thus must be a nutritional deficiency disease. Granted, some of his studies were a bit morally questionable, relying as they did on orphans, inmates, and patients in mental asylums. (This was often the case in early medical studies—in fact, it's hard to imagine how we'd have learned much of what we know about medicine or nutrition if there'd been such a thing as informed consent.) But even so, Goldberger's results should have been convincing.

When critics instead continued to insist that pellagra was caused by a microbe, Goldberger took an even more dramatic step: in 1916—around the same time that vitamin A was identified—he began hosting "filth parties" (his term!), in which he and fifteen other volunteers exposed themselves to bodily secretions of people sick with pellagra in order to prove that it was not a contagious disease. The volunteers, who were mostly fellow medical men, injected themselves with pellagrins' blood, swabbed pellagrins' nasal and throat secretions onto their own mucous membranes, and ate pills made out of pellagrins' urine, faeces, and dried skin flakes. "Measured quantities of the materials mentioned were worked up with cracker crumbs and a little flour into a pilular [*sic*] mass," wrote Goldberger, in a description of what sounds a bit like a recipe for the world's most disgusting hors d'oeuvre. Even Goldberger's wife, Mary, insisted on participating in the experiments. While the men wouldn't allow her to swallow the pills, she did persuade her husband to inject into her stomach seven cubic centimeters of blood from people sick with pellagra. The attending nurse, convinced that this was a suicidal act, broke into tears.

None of Goldberger's volunteers got pellagra. But despite this and subsequent experiments, politicians and scientists remained convinced that pellagra was caused by a germ. Their critiques of Goldberger could be quite flamboyant: at a meeting of the Southern Medical Association in November 1916, Alabama doctor J. F. Yarbrough claimed that Goldberger's "advice to discard all drugs and other means other than diet has cast a pall of gloom over our fair Southland and our cemeteries are blooming as do fields of grain after beneficent summer showers." But despite Yarbrough's accusation that doctors who followed Goldberger's suggestions were "crucifying their patients upon a cross of error," it was pellagra itself that continued to kill: ten thousand people in the United States died of it in 1929 alone.

The deaths continued until 1937, eight years after Goldberger's own death from cancer, when vitamin B3/niacin—which is found in foods including chicken, beef, whole-grain products, legumes, brewer's yeast, and avocados— was finally isolated and pronounced as pellagra's cure.[1] While Goldberger had been instrumental in pushing the idea that pellagra was a nutritional deficiency disease, he was one step away from discovering its true cause: he believed the mystery substance, which he'd given the unfortunate nickname

---

[1] The disease was also acknowledged as being triggered by diets that rely heavily on corn. Corn is problematic in two ways: First, it doesn't have much niacin, and what niacin it does have is bound to glucose and protein in the corn and isn't readily available (you can make it more available if you prepare it with an alkaline substance, as is done when making tortillas—which could explain why pellagra was not typically a problem in Mexico, despite its dependence on corn). Second, while it's actually possible for humans to synthesize niacin from the essential amino acid tryptophan, corn doesn't have any tryptophan in it. Many protein-rich foods do contain tryptophan, however, which explains why a protein-rich diet could often prevent or cure pellagra.

"P-P" factor (short for pellagra-preventive factor), was tryptophan, an amino acid that we now know our bodies can use to synthesize niacin.

The *New York Times* heralded the identification of niacin as pellagra's cure. "What this success means, the statistics proclaim adequately enough," said an article in 1938 titled "Authorities Sure of Pellagra Cure." "So far as the United States Public Health Service can determine 400,000 people succumb to pellagra in this country every year—an underestimate. If the diet is not corrected the death rate is as high as 69 percent. Worse still the mind is affected. . . . To restore the victims to health of body and mind by adding the proper food doses of a cheap chemical seems miraculous." By 1941, niacin's importance was so universally recognized that the American government temporarily mandated that it be added to store-bought bread.

The fact that it took so long for pellagra to be acknowledged as a vitamin deficiency disease becomes even more surprising when you consider that, by the time of Goldberger's death, vitamins' existence—and the idea that there were "miraculous" chemicals in food that were able to restore "health of body and mind"—had been embraced by both scientists and the public. The story of how that happened brings us to perhaps the most important moment in the history of vitamins: how the word "vitamin" itself, so familiar to us today, so trusted and adored, actually came to be.

Despite their initial reluctance to accept the idea that beriberi was a nutritional deficiency disease, scientists had eventually accepted the idea that rice polishings contained an "anti-beriberi factor," as the Dutch researcher Eijkman had phrased it, and began work to figure out what that substance was. One of these scientists was a Polish biochemist named Casimir Funk, a man whose name is inextricably linked with vitamins—not because he isolated any (he did not), but because of an accomplishment that was arguably even more consequential: he came up with the word.

The son of a dermatologist, Funk was born in 1884 and lived a peripatetic life that took him from Poland, to Switzerland, to Paris, to Germany, to London, to New York, back to Poland, to Brussels, to Paris, and back to New York again. It was an exhausting path, but one that also made him fluent in Polish, Russian, French, German, and English. By 1911, he was publishing most of his scientific work in English, which helped to prevent his writings from being lost to the English-speaking medical community—as had Kanchiro Takaki's and those of the Dutch researchers—because of the lack of a translation.

In the autumn of 1910, Funk joined the Lister Institute of Preventive Medicine in London, and was assigned the task of trying to isolate the substance in rice polishings that prevented beriberi. Working with colleagues, Funk figured out that the mystery substance wasn't an amino acid (his boss's pet hypothesis), and he also disproved the theory that it was caused by a toxin in the polished white rice. Next, he began to fractionate the rice polishings, using a series of chemical reactions to try to isolate the precise substance that was curing the birds. And I mean a *small* fraction—we now know that a ton of rice bran contains only about a teaspoon's worth of pure thiamin.

Eventually, Funk was able to isolate a few tiny samples of a crystalline substance that—at least in a couple of cases—cured pigeons with polyneuritis. So far, so good. In December 1911, he published a paper in the *Journal of Physiology* stating that avian polyneuritis was caused by a lack of an essential substance in the birds' diet that was necessary in only tiny amounts. The substance, he guessed, was an amine—a type of nitrogen-containing organic compound. He also suspected that there might be other substances like it.

Funk never completely isolated thiamin; it was later determined that his crystalline material was actually mostly niacin—Goldberger's pellagra-preventive factor, which rice polishings also contain—contaminated with a bit of thiamin. (Pure thiamin wasn't isolated till 1926 by Dutch scientists in Java; it took them 700 pounds of rice polishings to get 100 milligrams of thiamin crystals.) But the impurity of his substance ended up being less important than what Funk wanted to call it. Recognizing that the crystals, by preventing beriberi, were essential for life, he took the Latin word for life—*vita*—and combined it with the term for what he believed would be the common chemical structure of other similar yet-to-be-discovered molecules—"amine." *Vitamine*. It was the first public use of the word.

No one got to see it, however, because the Lister Institute's staff and the editorial board of the journal didn't approve of Funk's creativity. Instead, they gave his paper, which was published in 1911, the title "On the Chemical Nature of the Substance Which Cures Polyneuritis in Birds When Subjected to a Diet of Polished Rice," and called the mystery compound a "curative substance" rather than a vitamine.

Funk tried and failed a few more times to get his new word published.

Then in 1912, he had his opening: he was invited to write a review for the Lister Institute's publication, the *Journal of State Medicine*, about nutritional deficiency diseases—an article for which he didn't need to get the approval of the Lister Institute's staff. And so in June 1912, the word "vitamine" first appeared in print.

The paper, titled "The Etiology of the Deficiency Diseases" (subtitled "Beriberi, Polyneuritis in Birds, Epidemic, Dropsy, Scurvy, Experimental Scurvy in Animals, Infantile Scurvy, Ship Beri-Beri, Pellagra") was revolutionary. To start with, Funk proposed that beriberi, scurvy, pellagra, and rickets all "present certain characters which justify their inclusion in one group, called *deficiency diseases* ... caused by a deficiency of some essential substances in the food." (Emphasis mine.) It was one of the first times anyone had combined all four of these diseases under the banner of nutritional deficiencies. But Funk's most famous statement is undoubtedly this:

"It is now known that all these diseases ... can be prevented or cured by the addition of certain preventative substances," he wrote. "[T]he deficient substances, which are of the nature of organic bases, we will call 'vitamines'; and we will speak of a beri-beri or scurvy vitamine, which means, a substance preventing the special disease." Funk then went on to use the word "vitamine" repeatedly throughout his twenty-seven-page article, dropping it casually into his discussion as if it hadn't made its print debut just several pages before.

Funk's bold semantic move wouldn't surprise anyone who read his paper today—the word feels natural, as part of our nutritional vocabulary as "calorie" or "protein." But Funk's scientific peers weren't quite so quick to accept his term. At the time, no one had yet proved that all four of the diseases named by Funk were caused by nutritional deficiencies. Also, he hadn't actually isolated the substance he was writing about. What's more, there was professional rivalry going on, with Funk's scientific peers offering their own alternative terms—a collection of suggestions that included "accessory food factor," "nutramine," "food hormone," "fat-soluble A," and "water-soluble B."

Subsequent researchers were also wary of Funk's word because, while Funk did end up being correct that all four diseases were caused by nutritional deficiencies, the substances that cured them were not all amines—the nitrogen-containing organic compounds that originally gave Funk the "amine" part of vitamine. This meant that the word (at least as it was originally spelled) was chemically misleading.[2]

Nonetheless, by 1920, four of the substances we now call vitamins had been identified (if not chemically isolated), and a decision had to be made about what to call them. While the term "vitamine" was known by scientists and beginning to be used by food marketers, the four substances themselves were also often identified by letters: A, B (which at that point referred to only one substance), C, and D. To clarify things, the British biochemist Jack Drummond proposed in 1920 that the final "e" of "vitamine" be dropped in acknowledgment that not all vitamins were amines. He also suggested that scientists do away with all the "somewhat cumbrous nomenclature" in use at the time, and just call them "vitamin A," "vitamin B," et cetera.

Drummond's suggestion is still obviously in use today. But at the time when he proposed it, many early researchers, aware that the substances in question didn't fit into one chemical family, didn't think that it was going to last. Elmer McCollum, the American chemist who had coined the term "fat-soluble A," thought of "vitamin" as a placeholder that would "automatically fall into disuse when we come to possess definite knowledge of their chemical nature." The American physiological chemist Russell Chittenden later proposed that the word would "soon join the 'musty company of phlogistic, humors, animalcules, and kindred antiquated terms.'" At first, "this lettering of unknown, quasi-mysterious substances did much to popularize them and to make the world vitamin-conscious," he wrote. But "[t]here is no longer any scientific basis for maintaining such widely different chemical substances . . . under the same heading except for historical purposes."

---

[2] Also, while it's true that all the diseases Funk mentioned could be caused by deficiencies in the diet, later researchers discovered that the curative substances don't always need to come from food. Vitamin D, which is naturally present in very few foods, is created in our skin with the help of ultraviolet rays from the sun. If you have enough of the essential amino acid tryptophan, you can probably produce enough niacin to prevent outright deficiency. We can transform beta-carotene, the substance that gives carrots their orange colour, into clear yellow vitamin A (making beta-carotene what's known as a provitamin). Vitamin K and biotin can be produced by microbes in our gut. What's more, even the number of compounds defined as human vitamins can fluctuate: choline is usually considered a "conditionally essential nutrient," meaning that it's necessary for your diet only if your body doesn't have the chemical ingredients from which to make it. But, like an asteroid granted planetary status, it's occasionally mentioned as a fourteenth human vitamin.

These statements speak to one of the oddest things about vitamins: chemically speaking, there's no precise definition of what a vitamin actually is. But despite this lack of specificity, the word "vitamin" obviously did not fall into disuse—far from it. Instead, it has taken on a life of its own, used and abused by advertisers to such an extent that it can be seen as one of the most brilliant marketing terms of all time. Casimir Funk's biographer proclaimed that "the very term is pregnant with meaning"; as Funk himself readily acknowledged, "vitamin" wasn't just better than its competitors, it was brilliant, a linguistic concoction so evocative, so satisfying, that it "served as a catchword that meant something even to the uninitiated."

And it's true. The word carries a sense of both necessity and aspiration, the prevention of disease plus the potential for perfect health. Today, just over one hundred years since it was coined, "vitamin" has transcended its scientific roots, and grown more seductive than Funk himself could ever have imagined. Our bodies may depend on thirteen essential dietary chemicals. But it's *vitamins* that we're obsessed with.

4

The Journey into Food

It is abundantly clear that before the last century closed,
there was already ample evidence available to show that
the needs of nutrition could not be adequately defined in
terms of calories, proteins, and salts alone.
—FREDERICK GOWLAND HOPKINS,
NOBEL PRIZE ACCEPTANCE SPEECH, 1929

t's nearly impossible to open a magazine or go grocery shopping without
hearing about a new dietary chemical that's supposed to make us healthier.
Carotenoids, catechins, curcumin—we often don't have any idea what these
chemicals actually are or how they're supposed to work or even how they're
pronounced. But since "experts" have told us that they're good for us, we
incorporate these new terms into our everyday vocabularies and begin buying
products that claim to contain them.

This strange habit is largely vitamins' fault. By introducing chemical terms
like niacin and thiamin to the public, vitamins laid the groundwork for the
surprisingly chemical-oriented way we talk about food, with advertisements for
ketchup highlighting lycopene instead of tomatoes, and resveratrol appearing
in articles about red wine. But how did vitamins themselves come to be discovered? The fact that the word had been coined and vitamins' existence
suspected didn't mean anyone had identified a vitamin or knew precisely where
one could be found. In order for that to occur, scientists first had to recognize one of the basic tenets of nutrition: that food could be broken into parts.

That realization was pioneered by a cast of nineteenth- and early-twentieth-
century scientists who were committed to understanding and recreating the
chemical composition of food. These scientists coined much of the basic vocabulary we use to talk about nutrition, words that seem so familiar to us today that
we never think to wonder where they came from. What we don't realize, however, is that many of these scientists' questions still have not been fully answered.
This makes understanding their process—and the ways they came to terms with

their own ignorance—particularly useful for us to keep in mind as we grapple with our current chemical vocabulary and make daily decisions about what to eat.

By now, not only do we all know that food contains energy, but we are obsessed with measuring it: the concept of a calorie is so integral to our philosophy toward nutrition that it'd be easy to assume that the word "calorie" has existed for as long as humans have been having conversations about food. In reality, however, the first recorded use of the term occurred in 1825, and it had nothing to do with nutrition. Instead, the word appeared in a lecture about steam engines, and referred to the amount of energy required to raise the temperature of a kilogram of water by one degree centigrade. Though it's not how we usually think of it, that's still the definition of a calorie—which technically should be written with a capital "C"—today.

The connection between human metabolism and other heat-producing chemical reactions (including, eventually, steam engines) was first established by the eighteenth-century French chemist Antoine Lavoisier, a brilliant man whose research career was cut short, all too literally, when he was beheaded in the French Revolution. In one of his best-known experiments, Lavoisier put a guinea pig in a chamber surrounded by ice and measured the heat and carbon dioxide produced by the animal's exhalations. He then calculated precisely how much coal he needed to burn in that same ice chamber to match the carbon dioxide produced by the exhalations of that poor, chilly guinea pig and measured how much heat the lump of coal generated when burned. The amount of heat produced by the combustion of the coal was the same as the heat generated by the guinea pig, leading Lavoisier to correctly conclude that the metabolism of food in the body could be compared to a slow-burning fire. This realization also suggested a straightforward way to calculate how much energy a particular piece of food contained: you could burn the food and measure the resulting heat.

The adoption of the word "calorie" in the nineteenth century gave scientists a precise unit with which to put Lavoisier's guinea pig revelation to practical use, and in the late 1800s they began creating charts of the calorie counts—that is, energy content—of popular foods. They did so by placing samples of food in devices called bomb calorimeters—sealed containers surrounded by water—and then burning the food; the ensuing rise in the water's temperature revealed how many calories the food sample had contained. Today, few people would look at a 100-calorie Oreo snack pack and think, If I combusted those cookies, the resulting energy could raise a kilogram of water from its freezing point to its boiling point. But technically speaking, that's exactly what it means.

Of course, measuring the overall energy a food contains doesn't reveal the source of that energy, just as measuring a fire's heat won't necessarily tell you what's being burned. That next stage—namely, the discovery of the macronutrients and the ability to measure them individually—required the recognition that food wasn't just a solid lump of calories but instead could be broken into calorie-containing chemical parts. Not only did this perspective lay the groundwork for the eventual chemical isolation of vitamins, but it fundamentally changed the way our predecessors thought of nutrition—and how we still think of it today.

The concept of macronutrients was first proposed in 1827 by a British physician-turned-chemist named William Prout, who suggested that food contained three energy-providing "staminal principles," which we now know as carbohydrate, protein, and fat. A French physiologist, François Magendie, then proposed that perhaps Prout's staminal principles each had different purposes in the body, and in 1843 the great German chemist Justus von Liebig proclaimed in a book called *Animal Chemistry* that he had those purposes all figured out.

Liebig was an extremely influential chemist and oversize personality, whose accomplishments ranged from the recognition that nitrogen was an essential plant nutrient (leading to his development of the first nitrogen-based plant fertilizers) to the creation of what eventually became Oxo beef bouillon. Liebig never performed human nutritional experiments himself, but this did not stop him from coming up with theories based on what he had observed in plants. Since nitrogen is critical for plants, and protein is the only macronutrient to contain nitrogen, Liebig decided that protein must be the most important nutrient for humans (the word is derived from the Greek *proteios*, meaning "first quality"); he believed it was necessary for building and maintaining tissues and that it was the sole source of energy for our muscles. As for carbohydrate and fat, Liebig claimed that their calories could only be used to create body heat; they couldn't be used as fuel for muscular work. That's why Eskimos ate so much fat, he explained—they needed it to keep warm.

While Liebig was correct that our bodies need protein to build and maintain tissues, his theories on energy sources were not—it would be as if a car required one type of fuel to move forward and another type of fuel to keep its engine hot. We now know that carbohydrate is actually a common fuel source for our muscles (protein is used for energy if nothing else is available), and that body heat is produced regardless of which macronutrient is being burned. Whereas Liebig assumed that the forms of each macronutrient were interchangeable (that is, a protein is a protein, regardless of source), we also now know that protein, carbohydrate, and fat can themselves be broken down into subcategories (essential and nonessential amino acids, for example, or starch and sugar, or saturated fat and omega-3 fatty acids), and that these forms have different subtle effects in our bodies that we still don't entirely understand.

But despite their errors, Liebig's theories had a net positive effect: they helped jumpstart the field of nutritional chemistry, a result that he claimed, in the preface to *Animal Chemistry,* had been his goal in writing the book. "My object . . . has been to direct attention to the points of intersection of chemistry with physiology, and to point out those parts in which the sciences become, as it were, mixed up together," he wrote. "[I]f, among the results which I have developed or indicated in this work, one alone shall admit of useful application, I shall consider the object for which it was written fully attained. The path which has led to it shall open up other paths; and this I consider as the most important object to be gained."

New paths did indeed open up—in fact, the intersection of chemistry and physiology that he describes is what the study of nutrition is all about. And in the late 1880s, a Liebig-inspired cast of German nutritional scientists began pioneering work that both further established Germany as the nutritional research capital of the world, and eventually led to the discovery of the vitamins.

Among their many projects, the German scientists combusted foods in bomb calorimeters to quantify their calories. Then, with this knowledge in hand, they zoomed in a level deeper, using a technique called proximate analysis to determine these foods' chemical composition.

In proximate analysis—versions of which are still in use today—fat is extracted using ether or another solvent and then measured; protein is calculated based on the amount of nitrogen the sample contains. Carbohydrate is

whatever is left over. Proximate analysis can also be used to measure
food that are not burned for energy. To figure out how much water a
contained, for example, the scientists oven-dried the sample and then calcu-
lated the difference in weight; to measure its overall minerals—whose nutri-
tional purpose was still not understood—they combusted the food in a
furnace and measured the resulting ash. (Minerals, which are inorganic ele-
ments, don't contain carbon and therefore don't burn.)[1]

The work of these scientists becomes particularly impressive when you
realize that there are only five classes of essential nutrients—protein, carbo-
hydrate, fat, water, minerals, and vitamins—and they'd figured out how to
measure four of them. But these scientists couldn't capture vitamins. First,
vitamins contain carbon, which meant that they burned up in the bomb calor-
imeter along with the macronutrients. And even if the scientists had sus-
pected that vitamins existed, the equipment of the day wasn't sensitive
enough to detect them. The amounts of vitamins found naturally in foods are
so tiny that even today they're often quantified by indirect methods—say, by
measuring the growth of a bacterium known to be dependent on a particular
vitamin—rather than by trying to isolate and weigh them directly.

Unaware that they were missing anything, the German researchers used
the results of their analyses to create charts quantifying the amount of pro-
tein, fat, carbohydrate, water, and ash that various foods contained—
precursors to today's nutritional databases. These charts were primarily for
scientists, however, and the work was done in Germany; the average layper-
son had never heard of calories, let alone protein, fat, or carbohydrate. But that
was about to change. Thanks to the work of a chemist named Wilbur Olin
Atwater, we would never look at food the same way again.

If you have ever spent breakfast staring at the Nutrition Facts panel on your
cereal box, you have felt the influence of Wilbur Olin Atwater. Born in Johns-
burg, New York, in 1844, Atwater had a PhD from Yale in agricultural chemis-
try, which he'd continued to study—along with physiological chemistry—in

---

[1] Minerals only consist of one chemical—think calcium, or iron, or potassium. Vitamins,
on the other hand, are organic compounds, meaning that they contain carbon and are
made of more than one chemical (and are produced by living things).

Germany with some of the leading nutritional chemists of the time. Back in the United States, Atwater began evaluating the calories and macronutrients in hundreds of foods, and eventually arrived at a conclusion that today is common knowledge to anyone who's been on a calorie-restricted diet: protein and carbohydrate contain roughly four calories per gram, while fat contains nine.

Although Atwater's analytical work was undoubtedly valuable, the most influential aspect of his legacy was not his calculations or calorie counts, but his communications with the public. In 1887 and 1888, he published an influential series of articles in *Century* magazine called "The Chemistry of Foods and Nutrition" that encouraged readers to think of food not simply as nourishment, but as a sum of its parts. This represented a sea change in how Americans viewed food, which Atwater summarized in a later article from 1892—a quote that perfectly captures our current attitude toward nutrition:

"For the discussion I must take a different view of food from that to which we are accustomed," he wrote, "and consider, not the food as a whole, but the nutriment it actually contains, which is a very different thing. I must take account of its chemical composition, its nutritive ingredients and the ways in which they are used to nourish our bodies. I must talk, not of beef and bread and potatoes, but of protein, carbohydrates, and fat." He was proposing, in other words, the very reductionist attitude that is so dominant today.

But as is so often the case with reductionism, Atwater was missing something important. Inspired by Justus von Liebig, Atwater believed that the only two elements humans needed to keep track of were protein (for repair and maintenance) and calories—which by this point were considered interchangeable—to provide energy. Following this logic, the best diet would be the one that was both the cheapest and the most protein- and calorie-dense—which meant that Americans, with their affection for expensive cuts of meat and low-calorie, low-protein foods like vegetables, were spending more on their food than was nutritionally necessary. Atwater, who was interested in social betterment as well as nutrition, concluded that a solution to poverty was therefore staring society in the face: people could increase their disposable incomes if they chose their foods solely based on the calories and protein they contained. After all, every penny saved on food was a penny they could spend somewhere else.

To promote this idea, Atwater published tables that compared the calories in various foods; one pound of turnips, he claimed, contained only 139 calories compared with the 3,452 calories provided by the same amount of very fatty pork. Today, the judgment of most nutritionists would fall strongly on the side of the turnip, but to Atwater and his followers, the turnip—or any other low-calorie piece of produce—would clearly be inferior to a Jimmy Dean's pancake-wrapped sausage stick.

As a later vitamin researcher put it, "From the point of view of the peace of Atwater's soul, it was a very lucky thing that he never had to come face to face with men or animals whose nourishment was set up according to his directives. It must also be considered a piece of very good luck indeed that the homemakers of America made no attempt to feed their families according to his advice." Observers later commented that Atwater knew just enough to be dangerous, and we now know that if you followed Atwater's advice too closely, you might put yourself at risk of deficiencies.

But even though Atwater's nutritional advice was never followed en masse, his articles caused a revolution in the way the public thought about food, and eventually paved the way for its embrace of vitamins. For the first time, readers were introduced to the idea that they should be choosing foods based on their calories, carbohydrate, protein, and fat—a radical departure from the traditional view that food is a consolidated chunk of energy, the nutritional equivalent of a lump of coal. Indeed, it was such a new concept that even as late as 1915, nutritional scientist and fellow Voit student Graham Lusk commented that "[i]t has been said by some that they never will be converted to the belief that a knowledge of calories in nutrition is valuable." In this sense, reading Atwater's articles from his audience's point of view is like travelling back in time.

And yet, while it's hard to imagine not knowing about calories, Atwater's writing also seems strangely modern—probably because we've so wholeheartedly embraced the reductionist philosophy that he proposes. We buy foods based on the combination of chemicals and calories they contain. We debate ways to create social change through diet. We've accepted the notion that we should emphasize "health" over taste—even as, in true Atwater spirit, we continue to seek out cheap and calorie-dense foods. Though we're not consciously aware of it, Atwater comes along with us every time we visit the grocery store.

We have also largely bought into the arrogant assumption that underlies *all* diet recommendations, whether they're from Justus von Liebig, Wilbur Atwater, or the health gurus of today: the idea that we humans fully understand our nutritional needs. Taken to its extreme, this suggests that humans shouldn't have to rely on nature for nutrition at all; instead, it should be possible for us to use science—that is to say, chemistry—to reverse engineer food. On paper, this sounds both tidy and appealing, and has contributed to our acceptance of the highly processed and engineered products that currently line supermarket shelves. But just as we can't imagine nutrition without vitamins, future generations will likely marvel at our ignorance of things they take for granted. It's quite possible that, like Atwater, we know just enough to be dangerous. In nutrition, as is so often the case, hubris does not translate into health.

Oddly, by the time that Atwater was pushing his protein-heavy diet, some scientists had already recognized that man-made diets were nutritionally incomplete—even if they didn't know exactly what these purified diets, as they were called at the time, were missing. But many early twentieth-century nutritionists, Atwater included, either didn't know about or wilfully ignored the failures noted by these other researchers, perhaps reluctant to consider that nutrition wasn't quite as cut-and-dried as they hoped.

One person did publicly acknowledge them, however: a British biochemist named Frederick Gowland Hopkins, who's the next crucial character in vitamins' chronology. Today, Hopkins is remembered for his contributions in dismantling the idea that human-made purified diets were nutritionally perfect and suggesting the existence of what we now recognize as the vitamins.

Hopkins originally gained fame in 1901 when he discovered tryptophan, an essential amino acid, and managed to isolate it from protein. This was an exciting development not just because it underscored the relatively new idea that macronutrients could be broken into smaller parts, but also because it helped establish the idea that certain proteins—or, more specifically, amino acids—were "essential," meaning they couldn't be made by the body and were only obtainable from food.[2]

---

[2] Not only are our organs and tissues mostly made of protein, but as the building blocks for enzymes and hormones, proteins are also necessary to make our bodies run—in fact, one of DNA's best understood functions is as an instruction manual for making proteins. Each protein is made of a combination of molecules called amino acids (the so-called building blocks of protein) strung together in a specific order; this order dictates the protein's shape, structure, and function. Our cells can create some amino acids themselves, but others are essential, meaning we have to get them from our diet.

Then, on the evening of Wednesday, November 7, 1906—about five years before Casimir Funk first suggested the word "vitamine"—Hopkins gave a speech in which he touched upon the observation that eventually led to his receipt of the Nobel Prize, an award that he shared with the beriberi researcher Christiaan Eijkman.

Hopkins told the audience that during his amino acid experiments, he had noticed something odd: when he raised mice on supposedly complete purified diets, they failed to grow. It was an observation that, according to the logic of the time, didn't make sense; as long as the mice had enough energy and enough protein, they were supposed to be fine. Hopkins's discovery of tryptophan had already begun to poke holes in the idea that all proteins—and, for that matter, carbohydrates and fats—were interchangeable with each other. Now, caught once again between nutritional theory and the reality of his lab, Hopkins was left with a consuming question: What was missing from the mice's diets?

Hopkins hadn't yet done experiments to try to identify the absent substances that seemed necessary for growth, but he felt confident enough about their existence to draw a provocative conclusion, which he shared with his audience that night. It was an idea that he predicted "will one day become recognized as of great practical importance."

"No animal can live upon a mixture of pure protein, fat and carbohydrate," he announced. "The animal body is adjusted to live either upon plant tissues or the tissues of other animals, and these contain countless substances other than the proteins, carbohydrates, and fats." He continued, "In diseases such as rickets, and particularly in scurvy, we have had for long years knowledge of a dietetic factor, but though we now know how to benefit these conditions empirically, the real errors in the diet are to this day quite obscure. . . . [L]ater developments of the science of dietetics will deal with factors highly complex and at present unknown."

Just as the Dutch researchers had done when they suggested that specific substances in foods could prevent beriberi, Hopkins was referring to what we now know as vitamins. But unlike the Dutchmen (or, for that matter, Casimir Funk when he coined the word "vitamine"), Hopkins wasn't just talking about deficiency diseases; he was talking about the failure of human-engineered diets. If the results of his experiments were to be believed, then Liebig had been wrong: the fact that animals on these diets got sick meant that the scientific understanding of human nutrition must not be complete.

The post-lecture discussion ran so long that it took more than seven pages to summarize, and yet no one commented on Hopkins's bold nutritional claims. This was probably because his audience, the Society of Public Analysts, didn't have an explicit interest in nutrition. But despite Hopkins's later claim that it was "abundantly clear that before the last century closed, there was already ample evidence available to show that the needs of nutrition could not be adequately defined in terms of calories, proteins, and salts alone," few people in the nutritional world seemed to be paying attention to the idea he was proposing, either.

Hopkins himself later attributed this oversight to the early twentieth-century obsession with quantifying the energy in food, arguing that just as the discovery of disease-causing bacteria in the late 1800s had delayed the recognition of nutritional deficiency diseases, the enthusiasm for calorimetry and the macronutrients had made it difficult to imagine that other unknown nutritional factors might exist. But regardless of the reason, Hopkins, who had merely observed these effects—and wasn't aware of the various Dutch researchers' suggestive comments about rice-based diets and beriberi—chose not to say anything further on the subject until he had done specific experiments to identify these mystery substances. He was silent for five years.

Finally, in 1912, after a health breakdown and several years of further experiments (and around the same time as Funk's first successful publication of "vitamine"), Hopkins published his results. He had raised matching sets of rats, he explained, in strictly controlled environments. The only difference between the two groups was the "administration of a minute quantity of milk"—an amount so small that its solids made up no more than 1–4 percent of all the rats' food. According to prevailing logic, this shouldn't have made a difference. But, on the contrary, "the influence of the milk upon growth was so large," Hopkins wrote, "that it could not have been due to any alteration in the quality of the protein eaten or . . . in my own belief, to the presence of any known milk constituent." He attributed the difference to some substance— or substances—in the milk that were present only in tiny amounts, but which were essential for proper growth. Though he was aware of Funk's "vitamine," Hopkins did not use the word. Instead, he called these substances "accessory factors."

"[A]t first it seemed so unlikely!" he later explained in his 1922 acceptance speech for the Chandler Medal at Columbia University. "So much careful scientific work upon nutrition had been carried on for half a century and more—how could fundamentals have been missed? But, after a time, one said to oneself, 'Why not?'"

The *idea* of vitamins, in other words, was beginning to coalesce. But the word "vitamin" was known only to Funk's scientific readers, and no one had successfully identified a vitamin in food. While many people—Hopkins included—contributed to that process, one man played a particularly large role: Elmer Verner McCollum, the self-proclaimed "discoverer" of vitamin A.

Elmer McCollum was born to homesteaders in Kansas in 1879, and his first noteworthy interaction with vitamins supposedly occurred when he was only a baby. His mother had become pregnant while nursing him, and in an attempt to conserve her strength, she switched little Elmer to a boiled mixture of cow's milk and potatoes. The boiling was beneficial in that it killed some of the deadly microbes that were responsible for the era's high death rate of children raised on (contaminated) cow's milk. But the heat also destroyed the mixture's vitamin C.

According to his mother, by the time he was about ten months old, McCollum's skin broke out in brown spots and his joints became so swollen and sensitive that he screamed when she handled him. The community doctor decided that the reason his gums were bleeding was that he was having trouble teething, so he pulled out an unsanitized pocketknife and made cuts around the howling infant's gums. In one of his less earth-shattering hypotheses, McCollum suspected that this may have contributed to his later dental problems; either way, it certainly didn't cure what he was actually suffering from, which was infantile scurvy.

He likely would have died, but one afternoon—so the story goes—his mother held him on her lap while she was peeling apples. She offered him a few scrapings, and noticed how eagerly he ate them up. McCollum claimed that his mother believed that humans instinctively know what foods they need to eat, so when she saw her baby's enthusiasm for apples, she kept feeding them to him. He began to get better within three days.

Recovered, McCollum spent most of his early life barefoot on his family's farm, observing the details of farm life with an eye that, even then, hinted at his future as a scientist. In high school, McCollum somehow managed to balance his schoolwork with an afternoon job at the local paper and a night-time gig as a gas lamp lighter, all while still helping out on the farm. He made enough money to buy himself an entire set of the *Encyclopaedia Britannica*, which he read in his scarce spare time. Sleep deprived, too busy to socialize, and so overworked that his six-foot frame weighed scarcely more than 122 pounds, McCollum was somehow still elected class president in both his junior and senior years. He earned a BA and an MS from the University of Kansas; within several years he had also received a PhD in chemistry from Yale under the renowned nutritional researcher Thomas Osborne.

In 1906, the year after McCollum earned his PhD, his advisor in New Haven received a letter from the head of agricultural chemistry at the Wisconsin College of Agriculture, who was looking for someone trained in biochemistry to help with the follow-up to a groundbreaking experiment that had recently taken place.

The chief chemist of the agricultural research station had wanted to find out whether diets that contained an identical balance of macronutrients (which could be determined using the technique of proximate analysis) were actually nutritionally equivalent. So around the turn of the twentieth century, a few years before Frederick Gowland Hopkins's provocative speech to the chemical analysts, he had convinced the animal husbandry department to lend him two cows, which he then put on diets—one based on oats and the other on wheat—that had the exact same balance of carbohydrate, protein, and fat. If the diets were truly identical, as most scientists assumed them to be, then there should have been no effect on the cows. But rather than thrive, both cows' health began to deteriorate. Then the oat-fed cow died. Displeased, the chair of the animal husbandry department had taken back the other cow. Now, several years later, the head of the agricultural chemistry department had finally authorized a larger follow-up experiment to figure out why the cows had become sick—and he wanted McCollum to run it.

It's worth noting that the Agricultural Experiment Station at the University of Wisconsin was not a place associated with the cutting edge of human nutritional research. It was, after all, one of a group of agricultural research stations that were founded across the country as a result of the 1887 Hatch Act, which aimed to promote research into the fertility of America's soil, the development of new strains of plants and livestock, and the comparison of different types of feeds. Like all of America's agricultural research stations, the Wisconsin lab's mission was to improve agriculture, not human health.

What's more, the philosophy of most of these American agricultural research stations was actually quite German. The Hatch Act was German-inspired, and most agricultural scientists of the time believed in the updated version of Justus von Liebig's nutritional theories—namely, that as long as a diet contained adequate protein and calories, it didn't matter where they came from. According to this logic, the best animal feed was therefore simply the one that enabled farmers to get as much as possible out of their livestock while putting as little as possible in.

McCollum had no previous interest in nutrition—let alone livestock nutrition—and he was dissuaded by colleagues who "showed plainly that they thought I was making a mistake in casting my lot in investigations of nutrition." But he needed money and a steady job. And besides, the single-grain experiment "clearly indicated that something fundamental [about nutrition] remained to be discovered," McCollum later wrote. He wanted to figure out what that something might be.

The new version of the single-grain experiment that McCollum had been chosen to run was longer and more complicated than the original trial with the two cows. This one used sixteen heifers and ran for four years, starting when the cows were calves and following them for two gestations. The cows were broken into four groups, with each group receiving a diet that was "balanced" according to the reductionist science of proximate analysis. One was corn-based, one wheat-based, and one oat-based; as the control, the fourth group of cows received a diet that blended all three grains.

By the time Elmer McCollum arrived in 1907 (several years into the experiment), the effects of the different diets on the animals were already obvious. "They presented amazing contrasts," he later wrote. The wheat-fed cows had gone blind. Their calves were undersized and premature; none survived. The oat-fed cows managed to produce two calves that lived, but they were unhealthy, and the mixed-grain animals had similar results. Only the corn-fed cows gave birth to healthy offspring.[3]

---

[3] According to a 1997 article in the *Journal of Nutrition*, it would "seem highly probable

When the animals were bred a second time, the results were similar, except that by now two of the wheat-fed mothers were dead. Then, during the last year of the experiment, the researchers rotated the cows' diets. The cows that were switched to corn became healthier; the cows taken off the corn deteriorated. The researchers were left with a mystery. "Though all the cows had had feed of the same chemical composition, they differed enormously in physiological status," wrote McCollum. "I was employed to discover why it was so. It was a man-sized job for a beginner."

McCollum "pored over the journals of organic and biochemistry and sought out unusual tests to perform on the feeds, tissues, and excreta of the cows." He examined the cows' milk. He analyzed their blood and urine, and tried to figure out if there might be a toxin in the wheat. At first he was enthusiastic, sharing all of his new ideas with his bosses and colleagues, but as months passed, he became frustrated. The single-grain experiment had become well known by that point, with pictures of the cows and their calves making the rounds in agricultural colleges as far away as Europe. Yet McCollum himself lost confidence that he'd ever figure out what the difference in the grains might be. It was a problem with important implications, not just for agriculture, but for human nutrition as well: If scientists didn't fully understand the nutritional requirements of livestock—which these experiments were making abundantly clear that they didn't—then how could they possibly claim to understand those of people?

McCollum turned to the literature. Reading through back issues of a German publication called *Yearbook on the Progress of Animal Husbandry*, he noticed something that sparked his interest: between 1873 and 1906 there had been at least thirteen papers published on the failure of purified diets. (Much to his later dismay, however, he didn't come across the observations of the beriberi-studying Dutch researchers.) "I was struck by the fact that in every instance in which small animals had been restricted to such 'purified' diets they promptly failed in health, rapidly deteriorated physically, and lived

---

that differences in reproductive performance of the groups were due to differences in vitamin A status, owing to differences in carotenoid content of the rations [corn had more], possibly complicated in the case of the wheat-fed group by an inadequate intake of fat and a marginal intake of calcium."

only a few weeks," he later wrote. "I concluded that the most important problem in nutrition was to discover what was lacking in such diets."

By this point, however, McCollum was sick of working with cows. They were horrible research subjects: huge, expensive to feed, and with long life cycles. He wanted to use something smaller, cheaper, and shorter-lived—not to mention something that no one would care if he accidentally killed. He wanted to work with rats.

But McCollum's boss hated the idea, so McCollum spent two hours on a Saturday afternoon catching seventeen wild grey rats at an old horse barn at the research station, which he poured into a grain bag and smuggled into the basement of the Agricultural Hall. Then, when the barn rats turned out to be "too wild, too much alarmed, and too savage" to be good experimental subjects, he bought a dozen albino rats from a Chicago pet dealer to start his colony (rats were not yet common as research subjects). They cost him six dollars, for which he was never reimbursed.

Once he'd acquired his rodent research population, McCollum's first few experiments were "of a bungling nature," but he continued. Over the next few years, McCollum and his research assistant, a recent graduate of the University of California at Berkeley named Marguerite Davis, ran dozens of experiments on their rats.

Finally in 1912 (right around the time that the British biochemist Frederick Gowland Hopkins published his own rat/dairy product research), McCollum and Davis had what he called their "first great pioneering discovery." They'd found that if they added a tiny bit of butterfat or fat from egg yolks to a purified diet, the rats survived. But if the same diet was instead augmented with lard or olive oil, they died. Not only did this show that there was a difference between fats in general (a revolutionary idea whose health implications we're still trying to understand), but it demonstrated there was something particular in the dairy fats, necessary only in tiny amounts, that kept the rats alive. In 1913, McCollum and Davis published a paper that concluded that their results "strongly support[ed] the belief that there are certain accessory articles in certain food-stuffs which are essential for normal growth for extended periods."[4]

---

[4] Meanwhile, in New Haven, Connecticut, researchers Thomas Osborne and Lafayette Mendel—with whom McCollum had worked and studied, and who had started their

McCollum eventually determined that the rats' health also depended on a water-soluble substance, which they got via an alcoholic extract of wheat germ, and which appeared to be the same compound that prevented and cured beriberi. McCollum and his colleagues suspected that this extract contained the "vitamine" Funk had spent several years trying to isolate—a term that by then had become familiar to scientists, though not necessarily popular—but they didn't choose to use his word. This was partially because they didn't think the vital substances were all amines, and partially because they wanted to claim credit by assigning their own name.

Instead, McCollum suggested in a paper published in 1916 that the fat-soluble substance be called "fat-soluble A," and the water-soluble one "water-soluble B"—thereby not only explicitly rejecting Funk's word (along with Hopkins's "accessory factors"), but relegating Funk and his anti-beriberi substance, now known as B1 or thiamin, to alphabetical second place. McCollum's phrasing might not sound catchy, but it was a step up from his original suggestion: "unidentified dietary factor fat-soluble A."[5]

McCollum never actually isolated pure vitamin A—that wasn't accomplished till 1937—but he still gave himself credit for its discovery (he even went so far as to claim to have discovered vitamins, period). And, as we'll see in a bit, McCollum's "discovery" helped propel him to fame.

But what does it really mean to "discover" a vitamin anyway? It's not as if someone stumbled upon a vitamin growing in the woods, or found a pirate

---

own rat experiments in 1909—had reached a similar conclusion: there was a substance present in some types of animal fat, like milk, butter, and cod-liver oil, that was necessary for normal growth. Combined with McCollum's observations, this was a particularly surprising result—up to that point, scientists had assumed that all fats were essentially the same, and that their only function was as fuel. In 1913, the two teams published independent papers indicating that there was something in milk (Osborne and Mendel) and butter and eggs (McCollum and Davis) that was essential for rats' growth. The papers appeared in the same issue of the *Journal of Biological Chemistry*, but McCollum and Davis, who submitted their paper three weeks earlier than the other team, got most of the historical credit; Osborne and Mendel often end up in, well, a footnote.

[5] McCollum later discovered that "fat-soluble A" was actually two distinct substances, which we now know as A and D; he and subsequent scientists also realized that "water-soluble B" was a combination of chemical compounds (not all of which turned out to be vitamins) as well.

map labelled "X Marks the Vitamin C." Instead, for each of the thirteen vita-mins, "discovery" was actually a four-step process: the hypothesis that the vitamin existed (and in what foods), isolating it in pure form, figuring out its chemical structure, and eventually, learning how to synthesize it from scratch, a development that is necessary to produce enough for vitamin supplements and fortified foods. Each step in this process was highly competitive, since whoever succeeded first would likely receive historical credit—in fact, there were at least seven Nobel Prizes awarded specifically for vitamin-related work (though McCollum himself never received one). As soon as a vitamin was suspected, the race was on.

But as contentious as it is, the question of *who* discovered each of the vita-mins is ultimately less important than the fact that by 1920 the concept of vitamins had been embraced by scientists, and four of them—A, B (which we now know as B1 or thiamin), C, and D—had been given names. It didn't mat-ter that no vitamin had actually been chemically isolated, that the details of their chemical structures were not yet known, or that their purposes were not yet fully understood. Nutritional science had entered an entirely new frontier.

Government policy makers and researchers like McCollum were eager to publicize the new science of vitamins, both because they were proud of their discoveries, and because they wanted to translate them into practical dietary advice that would improve people's health. What they didn't antici-pate, however, was the enthusiasm that vitamins would arouse in food mar-keters, how easily this enthusiasm could be transferred to the public, and how quickly the word and concept of a vitamin would take on a life of its own. Not only did this lead to scientists' losing control of their own discov-eries, but it created a situation that still persists today, in which the scien-tific realities and limitations of vitamins are in perpetual conflict with our personal hopes and dreams.

5

From A to Zeitgeist

Hardly were the results of the laboratory experiments printed than the
new heroes—green vegetables, milk and oranges—were taken up with
the enthusiasm of a Lindbergh or a Babe Ruth. Lettuce, formerly a
vegetable Cinderella, within a decade occupied the centre of the
grocer's stalls, with exactly seven times its previous popularity. . . .
Oranges and apple drink stands sprang up on countless street
corners, and spinach was irrevocably installed in
the menu as childhood's major sorrow.
—EUNICE FULLER BARNARD,
"IN FOOD, ALSO, A NEW FASHION IS HERE,"
New York Times Magazine, 1930

A hundred years after Elmer McCollum's experiments, the word "vitamin" has become shorthand for health, its halo strong enough to sanc-
tify anything it touches. As a result, we act as if the mere presence of
vitamins in a food must mean that it is good for us, regardless of what else it
does or does not contain. Vitamins' amazing ability to prevent devastating
deficiencies has also helped to create our modern obsession with isolated
dietary chemicals—the assumption being that if food contains vitamins, then
it must contain other miraculous chemical compounds as well.

Invisible to the naked eye and typically tasteless, vitamins have left us
dependent on nutritional experts and product labels to tell us which vitamins
a food contains—and, by transitive, whether we should think of the food as
healthy. This mandatory outsourcing has primed us to accept the amazing
array of health claims, advertisements, and advice that we encounter each day.
As if that's not enough, we've also begun to credit vitamins with abilities that
go well beyond preventing deficiency, from curing colds and hangovers to
preventing autism and cancer—unproven assertions that strain common sense.
How did we get here?

I had a hunch as to what time period might hold the beginnings of the answer: the early 1920s, when vitamins leaped from the exclusive realm of scientists to the everyday lives of consumers, via stories in the popular press and the advertisements that accompanied them. I also had a hunch as to where I should look: women's magazines, since most were aimed at homemakers, and homemakers were largely in charge of choosing their families' food. So I headed to the periodicals room of the Free Library of Philadelphia and spent two days scanning through microfilms of every issue of *McCall's* magazine from 1922 to 1945.

All women's magazines of the time were packed with vitamin-related ads, but I was interested in *McCall's* in particular because I knew its editors had published several decades' worth of columns by Elmer McCollum. By the time he started writing his articles, McCollum—who liked to carry around photographs of malnourished cows and rats in his pocket like baby pictures to demonstrate vitamins' importance—had moved on from the University of Wisconsin; he was now head of the Department of Chemistry at the newly created School of Hygiene and Public Health at Johns Hopkins University, where he stayed till his retirement in 1946.

McCollum was one of several nutritional chemists to write articles for the popular press, and his column, which ran from 1923 till the mid-1940s, was the most enduring. Geared toward "the experienced homemakers on McCall Street" (the magazine's term for its readers), the column's purpose was to help homemakers put the new vitamin-enhanced understanding of nutritional chemistry to use in the kitchen. I wanted to see what advice and information he was passing along to American homemakers, since it was primarily through women that the "newer knowledge of nutrition," as McCollum himself referred to this vitamin-inspired approach to eating, was imparted to American families and eventually passed down to us.

When I finally finished amassing his columns, it was clear that McCollum was the perfect guide to this wild new nutritional land. Not only had he been at the very cutting edge of early vitamin research, but he was also able to write in an accessible style, with a down-to-earth folksiness that put readers at ease. As a result, it wasn't long before the six-foot-tall 120-something-pound bow-tie-wearing butter lover had become a nutritional celebrity.

And yet despite the fact that he was writing in what was arguably the most exciting time vitamins had ever known, McCollum soon encountered the same challenge faced by nutritional columnists today: how to write about the same topic every month in ways that seem exciting and fresh. ("You *already* wrote about canned milk," I told my microfilm reader at one point, putting me in the company of the high percentage of periodical reading room patrons muttering things to their newspapers.) Sometimes he veered off into pet topics, dental hygiene high among them. But most columns touched on the same nutritional message again and again: if you wanted to keep your family healthy, argued McCollum, you'd better eat your vegetables and drink your milk.

This recommended meal plan, which McCollum called the Protective Diet, was built around vitamins. Remembering the problems he'd observed in rats living on man-made purified diets, McCollum was concerned that Americans' increasing taste for refined and processed foods was putting them at risk for health problems, including those caused by vitamin deficiencies. So he pushed people to eat more of what he called "protective foods," including leafy greens like spinach, kale, collards, turnip and beet greens, and two daily salads that included raw fruits and vegetables.

Besides produce, one of McCollum's other favourite protective foods was whole milk—he advocated that Americans of all ages drink at least a quart daily. He pushed milk partly because of his allegiance to Wisconsin (a major dairy state) and the role it had played in his own research, and partly because milk, even unfortified, is a good source of calcium and several vitamins, including some of the Bs. (Most of the vitamin A and D currently in milk, particularly low-fat, are synthetic additions.)

The irony of McCollum's milk obsession—and indeed, of America's current dairy-heavy nutritional guidelines—is that some 65 percent of people in the world are lactose intolerant after childhood, including more than 90 percent of some East Asian populations and a high percentage of people of West African, Arab, Jewish, Greek, and Italian descent; the fact that only 5 percent of Northern Europeans are lactose intolerant stands out as a jarring exception. Most people lose the ability to make lactase, the enzyme necessary to break down the sugar in milk, after weaning.

But at the time, the most shocking element of the Protective Diet wasn't the milk; it was the produce. McCollum's emphasis on fruits and vegetables was a radical departure from the nutritional advice of the time, which was largely based on the calorie- and protein-heavy guidelines endorsed by early twentieth-century scientists like Wilbur Atwater, the chemist whose articles encouraged Americans to think of their foods in terms of calories, protein, carbohydrate, and fat. Now McCollum was suggesting that American homemakers step beyond calories and macronutrients into a realm of nutritional detail so recently discovered that scientists, including McCollum, were still struggling to chart it. Despite the fact that the word "vitamin" was barely a decade old—and scientists' understanding of the substances was even younger—homemakers were supposed to know that vitamin C prevents scurvy (a disease they'd likely never seen) and can be found in green peppers, and that vitamin A is important for the immune system and is present in eggs.

Perhaps unsurprisingly, McCollum's encouragement to think of food in terms of its barely understood chemistry often left readers both nervous and confused; in fact, vitamins were significant early contributors to the anxiety over nutrition that still haunts us today. To allay some of these concerns, *McCall's* ran a feature in February 1935 called "Are Your Menus Right?" that tried to illustrate, through the eyes of a young, starstruck *McCall's* editor, how to put McCollum's philosophy into practice. In addition to illustrating how much American homemakers were already worrying about vitamins, it is also an early example of our enduring desire for nutritional gurus who can tell us what exactly to eat.

"[M]y heart was pounding with anticipation, and (I may as well admit it!) with a trace of stage-fright," the editor recalls, describing her elevator ride up to his office. "Would my questions seem ridiculously simple to this distinguished scientist?"

Luckily, they appear not to be. "I'd been thinking of him for years as a personage," she continues breathlessly, "when all the time he is a person, a simple, friendly person who makes you feel instantly his pleasure in being able to help you." When she arrives at his office, she continues to fawn. "I don't know much about food chemistry, but I do understand the Protective Diet," she says to McCollum. "Most careful homemakers do, don't they?"

"Let me show you something," says McCollum, ignoring the potentially condescending implications of her word choice (*careful* homemakers?!) as he pulls a sample menu off his desk. "This woman says her family won't drink milk, but she feels she's made up for that by feeding them plenty of vegetables. What do you think of these meals?"

The editor studies the menus—breakfast is cereal with cream, toast, and bacon; "luncheon" is a vegetable plate (peas, carrots, baked potato), French pastry, and tea. She is aghast, horrified.

"Why, she's neglected *all* the protective foods!" she cries.

"Can you improve those menus?" asks McCollum.

"I'd like to try." She takes a pencil and spends ten minutes poring over the meal plans, adding some cabbage, a pineapple salad, and as much milk and eggs as she can manage. The toast is now egg-drenched French toast. The French pastry is swapped for milk-rich chocolate pudding. A cherry pie is now cream-filled coconut custard.

McCollum approves of this indulgent revamp. "Simple changes," he says, "but they turn a dangerously deficient diet into one that satisfies all the body's needs." He launches into a mini sermon. "Perhaps in talking of the Protective Diet, I should always remember to say that the Protective Diet protects our health—protects us from many common ailments which we are likely to accept as necessary evils—protects us from early aging—protects us, sometimes, from emotional intemperance."

Before he can elaborate, their conversation is interrupted by lunch ("We talked about everything—Alaska, tree crops, Emily Dickinson, prize contests, the Choctaws"), and the editor leaves feeling good about herself and her meal plans ("Did I feel proud? I came home positively strutting"). She also comes home convinced that "Dr. McCollum's famous laboratory needed to be closer to my kitchen—and all the other kitchens on McCall Street." Whether the women in charge of those kitchens were actually reassured by the article, however, is another question.

The idea that women's choices in the kitchen needed to be "officially" approved by nutritional scientists represented an important perception shift in who could be trusted to decide what Americans should eat—and the insecurity caused by this shift left the public vulnerable to any person or product that claimed to offer answers. But McCollum's supposed encouragements also hid an even more frightening assumption: that any un-careful homemaker—that is, any woman who didn't follow the nutritional advice of outside experts—was playing Russian roulette with her family's health. Like any truly successful health guru, McCollum wasn't just a cheerleader. He was also a fearmonger.

It was a role that in some ways he embraced. For example, when a confused homemaker wrote to McCollum in 1938 asking him for clarification ("Maybe you think I'm dumb—but I get all mixed up trying to remember which vitamins are in what!") he responded to her request for a "simple chart that will tell me the vitamin story at a glance" with an infographic that likely terrified her. Titled "A Vitamin Primer," it listed the known vitamins next to columns describing "What it does for you," "Where you find it in large amounts," and "What happens if you don't have enough." Not enough vitamin C? "Small surface blood vessels rupture. Teeth loosen, fall out and die." Too little vitamin D? "Rickets develop in children. Teeth are poorly formed and decay. Heart action is affected." Skimp on vitamin E, and fetuses "in embryo cease to develop; may be absorbed by mother's body."

Loose teeth? Reabsorbed embryos? I could barely finish reading the chart without popping a multivitamin. Granted, the deficiency symptoms that he described are technically true. What McCollum left out of his chart, however, is that these are the results of *severe* deficiencies, not just a missed salad here or there—and that these levels of deficiency were not, and are not, common problems in the United States. Granted, our diets today have much room for improvement. But then, as now, there was no mass outbreaks of scurvy or beriberi in America. Rickets was only a concern in certain populations of the urban poor, and American mothers weren't reabsorbing their embryos because of a lack of vitamin E. (The reabsorption effect has only been seen in rats that are deliberately vitamin-starved.) With the possible exception of pellagra, the niacin deficiency disease that was a true problem in the South at the time, much of the increasing concern about vitamins was founded on a fear of deficiencies that did not—at least in America—exist.

McCollum himself admitted that most Americans weren't suffering from overt vitamin deficiencies, but appeared to feel no qualms about raising the alarm; instead, he played into—and even encouraged—a fear of nutritional deficiency that's still with us today. "There are thousands of people with *tendencies* toward these troubles," he told the *American Magazine* in 1923. "The round shoulders, flat chests, and poor teeth seen so frequently in school children are in great measure the result of faulty diet. Owing to the same cause, we see many adults growing old prematurely and suffering from bad health, which shows itself in a great variety of ways—in their discouraged mental outlook as well as in their physical condition."

Moderate vitamin deficiencies, in other words, could still cause physical

and emotional problems even if the deficiencies were unquantifiable, subjec-
tive, and way beyond anything proven in the lab; just as a great jazz musician's
genius can hide in the notes he doesn't play, some of malnutrition's most
insidious dangers could lurk where obvious symptoms didn't exist.

But deficiencies in *what*? McCollum both implicitly and explicitly
attributed most of these problems to inadequate vitamins. Not only was this
pure speculation—at the time there was little data on Americans' food consumption and no easy way to measure vitamin levels in foods or in human
bodies—but it had an unintended semantic consequence that hugely affects
us today. By using "nutritional" deficiencies and "vitamin" deficiencies interchangeably, and by promulgating the idea that vitamins' powers extended far
beyond the prevention of specific deficiency diseases, McCollum helped to
establish the word "vitamin" as a synonym for "health." But while it's true
that a healthy diet must include vitamins, vitamins themselves do not define
a healthy diet. There are too many other important compounds in food. We
may interpret them as synonyms, but the two words do not actually mean the
same thing.

Regardless of whether Americans actually followed McCollum's specific
dietary recommendations, the public's growing awareness of vitamins—which
was also being bolstered by government educational campaigns about the
importance of fruits and vegetables—was instrumental in changing people's
attitudes toward food. "In the vegetable kingdom, as in the kingdom of
heaven, the last suddenly became first, and the first last," wrote journalist
Eunice Fuller Barnard in her 1930 article on the cultural shift. "And viands
like lettuce with scarce a calory [*sic*] to their names, became the *sine qua non*
of the therapeutically favoured diet." Vitamins were now an important metric
by which food was to be judged, and Americans of both genders embraced
the notion that *careful* homemakers had a responsibility to ensure that their
families—through food and, later, supplements—had enough of each. How
much was enough, though? Nobody knew.

Given the uncertainty that continues to surround the RDAs, it makes sense
that Elmer McCollum's *McCall's* readers felt lost—the letters he received from
his readers clearly communicated their confusion over vitamins and their desperate desire for a knowledgeable, expert guide. Their anxiety is palpable and

all too familiar, because we still experience the very same anxiety about food today.

But not everyone was—or is—so distressed by the unanswered questions about vitamins, nutrition, and the definition of a healthy diet. Instead, when I looked at the advertisements that ran alongside McCollum's articles, it was obvious that to food manufacturers and their advertising firms, vitamins were like a heaven-sent gift: tasteless, invisible, immeasurable substances that needed to be eaten in some unknown amount every single day.[1] No longer was advertisers' nutritional vocabulary limited to the "energy" or protein in food. Nor were people supposed to buy foods based on what they actually enjoyed. "The food manufacturers have discovered a new language," wrote one nutritional chemist in a 1929 issue of *Good Housekeeping*. "Old staples that you and I bought because we liked the taste and found them 'filling' are now appearing in the advertising pages with new appeals to attention. They're rich in vitamins! Apparently that statement ought to be enough to make us open our pocket-books and purchase forthwith."

Vitamin C "cannot be stored in the body longer than 24 hours," warned one ad for Sunkist lemons. "It is essential that it be replenished daily." Manufacturers of cod-liver oil began referring to it as "bottled sunshine" because of its vitamin D content, came up with a mint-flavoured version, and advertised its supposed ability to give babies "well shaped heads." Iceberg lettuce, which is essentially water in leaf form, became "Nature's Concentrated Sunshine"; bananas were a "Natural Vitality Food." Ralston Wheat Cereal put "the B1 in Breakfast." "New research" suggested it was probably a good idea to "start or end One Meal a Day with Canned Pineapple." If you didn't want to risk vitamin starvation ("a danger that gives no warning!") you'd better eat Del Monte "vitamin-protected" canned foods. Schlitz Sunshine Vitamin D Beer was launched in 1936 with the tagline "Beer is good for you . . . but SCHLITZ, with Sunshine Vitamin D, is *extra* good for you."

Marketers weren't just content to use vitamins to boost the appeal of already popular foods. Instead, once they had used vitamins to establish the

---

[1] Today, we have the ability to measure the amount of vitamins in foods (often by using high-performance liquid chromatography or measuring the growth of vitamin-dependent bacteria). But vitamins are still invisible and usually tasteless, and their presence continues to be exploited by food marketers.

concept that foods' value should be judged not by how they tasted, but by whether they'd been vetted by nutritional scientists, they moved on to exploit vitamins' true magic power: their ability to spur demand for products that no one would otherwise think—or want—to eat. We're still surrounded by examples of this (consider the successful marketing of the goji berry as a health food), but my favourite historical example of this phenomenon is actually a product—wildly popular in its time—that few people would recognize today.

Yeast cakes.

Yes, yeast cakes—that is, yeast that's been pressed into a cake and wrapped in foil like a bouillon cube. It's technically the same yeast that's in active dry yeast, which is a fungus called *Saccharomyces cerevisiae* (the name roughly translates as "sugar fungus of beer") that's used in baking and beer brewing. *S. cerevisiae* feeds upon fermentable sugars and releases ethanol and carbon dioxide as waste products. The carbon dioxide bubbles are what cause dough to rise; the ethanol explains Fleischmann's gin. But active dry yeast is, well, dry; yeast cakes, which are made from fresh or "compressed" yeast, are not. Their moisture makes them go bad quickly, which is the main reason that fresh yeast was mostly supplanted in the 1940s by the less perishable granular powder that's common today.

Fresh yeast is still used frequently by commercial bakers, but it's no longer a household term. In the 1920s and 1930s, however, everyone knew about yeast cakes, because they were one of the first vitamin-inspired food fads. As is true for most yeast, fresh yeast is an excellent source of B vitamins, which the yeast produces itself. And if you irradiate it, yeast will also produce vitamin D via a process similar to what occurs when sunlight hits our skin.[2] Thanks to a massive "Yeast for Health" advertising campaign by Fleischmann's that emphasized their vitamin content, yeast cakes became such a popular, faddish health food that they were available everywhere from groceries to cafeterias, lunch counters, and soda fountains—offered specifically so that consumers, who ate them whole or dissolved in drinks, could get an extra vitamin fix wherever food was sold.

The first ads claimed only that yeast cakes were full of B vitamins and could treat constipation (both of which were true), but it wasn't long before the

---

[2] A nutritional note for vegans: Yeast does not naturally produce vitamin B12. If you are trying to get B12 via yeast, use nutritional yeast that's been fortified with it.

claims for yeast cakes—which, by the way, taste disgusting—began to multi-
ply like the yeast spores themselves. By around 1920, yeast cakes were being
advertised as a fix for skin troubles, stomach issues, and a "general run-down
condition." You should eat a cake before every meal—but be patient: results,
the ads emphasized, might take months to see.

The campaign was enormously successful: Fleischmann's yeast sales tri-
pled between 1917 and 1924. Between 1924 and 1925, the company's net
income was up an additional 75 percent, and by 1927, Fleischmann's sale of
yeast in the United States was 2.45 pounds *per capita*. As the ad company that
had created the campaign put it in an internal report, "The effect of the 'Yeast
for Health' campaign in increasing sales of Fleischmann's Yeast is very clearly
shown."

As more vitamins were discovered and new fortification techniques were
developed, the nutritional marketing possibilities for yeast continued to grow.
In late 1929, Fleischmann's began irradiating yeast to create vitamin D, supposedly to fight against what ads called a "sun-starved" race—"soft-boned,
weak-muscled, teeth a prey to decay." Once vitamin B had been divided into
thiamin (B1) and riboflavin (B2, then known as vitamin G), both of which
yeast naturally produces, the ads were able to claim it was "rich in *three* vitamins." Then in 1934, a new strain of yeast was introduced ("'XR' Yeast—
that's the scientists' name for it," said one ad) that supposedly contained
vitamin A. According to the ad copy, this meant that yeast could reduce colds.
But don't worry, a parenthetical assured traditional yeast users, "this new
Fleischmann's yeast is as good as ever for baking."

By the mid-1930s, yeast's four vitamins and unspecified "minerals and
hormone-like substances" had emboldened its advertisers to the point where
they actually began to argue that yeast was better for health than straight-up
fruits and vegetables. Borrowing Elmer McCollum's Protective Diet terminology, an ad in 1935 claimed that Fleischmann's yeast "supplies 'Protective
Substances' your stomach [and] bowels need to work properly," and that "[n]o
other food, even fruits and vegetables, gives you enough of them!"

Eventually, the ads were too much. The Federal Trade Commission filed a
cease-and-desist letter against Fleischmann's parent company, Standard
Brands, in 1931 protesting what it claimed were misleading ads; seven years and
many claims later, the FTC finally succeeded and Fleischmann's was forced to
cut back on its advertisements' assertions. By then, the ads had included claims

that yeast and its accompanying vitamins would prevent tooth decay; tone and strengthen your intestinal muscles; prevent pimples, "furry tongue," and colds; cure "fallen stomach"; improve your breath; cure depression; reduce headaches and fatigue; give you "pep"; eliminate crying spells; help your digestion; clear poisons from your system; raise your skin's "self-disinfecting power"; sharpen your intellect; and prevent you from becoming fat. One ad from 1937 included a testimonial claiming that Fleischmann's yeast, whose abilities apparently now rivalled those of Jesus, had restored a woman's ability to walk. The era of using vitamins to spark food fads had begun.

By 1941, McCollum's articles and food companies' aggressive advertising campaigns—yeast included—had already carried vitamins from scientists' labs to family kitchens, firmly establishing them as part of America's popular culture. Now, America's impending involvement in World War II was about to take them even further, elevating vitamins from a domestic concern managed by homemakers to a matter of national defense.

On May 26, 1941, several months before Japan's attack on Pearl Harbor thrust the United States into the war, some nine hundred men and women from across the United States—an assorted mix of home economists, nutritionists, physicians, social workers, public health officers, farm organizers, and representatives from consumer groups—gathered at the grand Mayflower Hotel in Washington, DC. Just a day later, spurred by fears of the Nazis' increasingly obvious goal of world domination, President Franklin Delano Roosevelt would declare an unlimited national emergency. But these delegates hadn't assembled to discuss bombs or battle plans. They had come to talk about food.

It was the National Nutrition Conference for Defense, a three-day gathering that marked the next phase of vitamins' spread. The goal of the conference, wrote President Roosevelt, was to "explore and define our nutrition problems, and to map out recommendations for an immediate program of action." He told delegates that he considered food and food policy to be an integral part of national security; the conference's final report affirmed that "[t]o neglect food would be as hazardous as to neglect military preparedness."

As the conference's speakers repeatedly stressed, nutrition and, more specifically, vitamins were pressing issues that would be critical to America's chances of victory. By the beginning of World War II, many military and nutritional experts had become convinced that millions of civilians and soldiers—one conference speaker estimated the number at 75 percent of all Americans—were suffering from deficiencies in vitamins and minerals that would leave the nation both physically and mentally unprepared for war.

According to historian Harvey Levenstein, the people who made these claims were "undaunted by mortality statistics that seemed to show quite the opposite"; indeed, "thanks in large part to the conquest of pellagra," deaths from vitamin deficiency diseases "had plummeted into relative insignificance by the end of the [1930s]" and "evidence of a link between these [supposed] deficiencies and poor health" was missing. Nonetheless, claimed the speakers, the stakes were high. If Americans were starved for micronutrients, America might not win the war. And if America lost? "[O]ur way of life will fail," conference chairman Paul McNutt predicted, "perhaps forever."

Food had been a concern in World War I as well, when the popular slogan "Food Will Win the War" expressed the government's concern that every American have sufficient calories to eat (as made sense, given the belief at the time that food's sole purpose was to provide energy and protein). But as a 1941 article in the *New York Times* argued, for World War II, the word "Food" should be swapped with "Vitamins." Now the goal of nutrition was not merely caloric sufficiency, but nutritional optimization, giving all Americans access to the micronutrients thought necessary to maintain complete and total health. Micronutrient optimization was arguably an even more daunting task than preventing starvation, though, because the goal itself was undefined: then, as is still true today, there was no consensus on what "optimal health" actually meant, let alone an agreed-upon way to achieve it.

Nonetheless, there was agreement that "optimization" referred in part to maximizing Americans' physical abilities—as the same article in the *Times* put it, "You cannot put into heavy industry a man who has been subsisting on a deficient diet for ten years and get anything out of him." Supposedly, vitamins could help: the conference chairman spoke admiringly about a truck company that he claimed had reduced its rate of accidents at night by giving its drivers bags of raw carrots at the beginning of every trip, and described a gunner in Britain's Royal Air Force with an "extraordinary record in nailing Nazi aircraft in the darkness" whose friends called him Carrots "because he was constantly munching that succulent root." (In reality, the Royal Air Force started a rumour that it was feeding carrots to its night pilots to hide from the Germans the true reason for their improved accuracy in the dark: radar.)

But consensus was also gathering around the idea that inadequate nutrition—including vitamin deficiencies—could affect not just people's physical abilities, but their personalities. Food companies had been pushing this idea for much of the previous two decades—as far back as 1927, Grape-Nuts had run an ad suggesting that poor nutrition could put children at risk of "unfortunate personality traits" including self-centredness, shyness, lack of confidence, selfishness, jealousy, depression, and self-pity.

Now, however, it wasn't just food marketers making these assertions: in 1942, the chemist Roger Williams proclaimed in his acceptance speech for Columbia University's prestigious Chandler Medal that "[t]here can be no doubt that much dullness on the part of school children, particularly among the lower-income groups, can be traced in part to a lack of the proper kind of food and specifically to the lack of enough vitamins. . . . Since an ample supply of vitamins can foster a higher intelligence in human subjects, it also has the capability of fostering morality." Vitamins weren't just essential for physical health; they were taking on a mental and moral dimension as well.

The National Nutrition Conference for Defense's final recommendations, which were presented directly to President Roosevelt, went even further: "There seems no reason to doubt," they said, "[that] by the use of the modern knowledge of nutrition we can build a better and a stronger race, with greater average resistance to disease, greater average length of life, and greater average mental powers." Correcting Americans' supposed nutritional deficiencies could, in other words, improve the definition of an American.

There is, one might note, a certain irony in the idea that America's leaders were talking about race building while simultaneously waging war against the Nazis. But creating a stronger race through nutrition was one of the very things Americans feared the Germans were trying to do. Eleanor Roosevelt, who delivered the conference's closing speech, spoke of ominous reports she had heard of the calmness and energy of young German men, which she believed came "very largely from the health they have built up in young people; what they have given them as they grew up through proper nutrition and proper surroundings." Three months after the nutrition conference, the *New*

York Times reported that Hitler had established a special institute for vitamin research, that all German margarine was being fortified with vitamin A, and that "a systematic effort has been undertaken by the German health and food authorities to make up for any deficiency by supplying certain groups of the population with synthetic vitamin pills or drops."

Dr. Russell Wilder, chairman of the National Research Council's Food and Nutrition Board, went so far as to claim that in addition to starving them, the Nazis were deliberately restricting the vitamins in the diets of the people they conquered in order to reduce them "to a state of mental weakness and depression and despair which will make them easier to hold in subjection." Later, its infeasibility notwithstanding, rumours flew that the Nazis were actually *destroying* the vitamins in defeated nations' foods. Vitamins weren't just a matter of national security; they were a matter of national character. And with so much at stake, the battle to ensure that Americans had adequate amounts of them needed to be actively fought—and won.

The question was how to do so. "Our problem," said Wilder, "is to reach Mrs. Tom Jones in terms she can translate into today's dinner. Milligrams and riboflavin naturally do not mean anything to her."

One could argue that by that point, the articles written by McCollum and his peers had ensured that milligrams and riboflavin *did* mean something to Mrs. Tom Jones, even if no one could tell her exactly how many milligrams of riboflavin her family actually needed. But regardless, the American government soon began to join food marketers, Elmer McCollum, and other popular nutritional writers in spreading the gospel of vitamins from scientists' labs to American kitchens. The result was a wave of new educational efforts, most notably a 1943 campaign, sponsored by the US Department of Agriculture, called the Basic Seven.

The campaign's most famous poster showed an illustrated pie chart with surprisingly nuanced, and hard-to-remember, distinctions. "Butter and fortified margarine" were separate from "milk and milk products"; "oranges, tomatoes and grapefruit" were separate from both "green and yellow vegetables" and "potatoes and other vegetables and fruits." The poster's tagline, "In addition to the basic 7, eat any other foods you want," likely didn't help Mrs. Jones's confusion, either. Nonetheless, the publicity of the Basic Seven campaign (which was quickly embraced by food advertisers and used as proof of the nutritional value of particular products) further raised Americans' concerns about the adequacy of their diets. The Basic Seven eventually evolved into the USDA's modern Food Guide Pyramid and MyPlate nutritional campaigns—another example of how America's World War II fears of vitamin deficiency still affect the way we appraise our diets today.

From a marketer's perspective, the ultimate hope for a food ingredient is that it will transcend mere trendiness to become part of the cultural zeitgeist, something considered so desirable, so indispensable (and ideally so imperceptible) that consumers will be attracted to any product that contains it. In the late 1930s and early 1940s, one vitamin in particular made the leap. And while this vitamin no longer holds the public's attention, it set the stage for a story that continues to repeat itself through a rotating selection of vitamins, minerals, and other trendy dietary chemicals.

It made an appearance on the first night of that 1941 National Nutrition Conference for Defense when, after a long day of meetings, US vice president Henry Wallace took the stage. Fifty-two years old at the time, with a side part that lurched into a stiff crest of hair, he had dabbled in Zoroastrianism and was described by a colleague as "a person answering calls the rest of us don't hear." When Wallace ran for president with the Progressive Party in 1947, one writer described his failed candidacy as "the closest the Soviet Union ever came to actually choosing a president of the United States." (He was more respected in his role as secretary of agriculture, and now has a USDA agricultural research centre named after him.)

Despite his reputation, Wallace's message that night was serious and specific. He told the audience that he had recently been listening to a radio show whose announcer had asked a question that resonated with Wallace, as he, like the conference's other delegates, reflected on what role nutrition could play in national defense. "What is it that puts the sparkle in your eye, the spring in your step, the zip in your soul?" the announcer had asked.

Wallace was already a fan of vitamins, having previously credited them with providing what he called "that feeling of 'health-plus.'" Now, once he'd recognized what the announcer was talking about—namely, his "old friend, vitamin B1"—the connection between thiamin and Americans' well-being seemed obvious: "It does seem that in the diet of a great many people in the United States," the vice president continued, "the addition of the different types of vitamin B makes life seem tremendously worth living."

Thiamin, as you may recall, is the vitamin—found in rice husks and other whole grains (and yes, yeast)—that prevents the leg swelling and cardiovascular nutritional deficiency disease of beriberi. The first water-soluble vitamin to be isolated, its molecular structure had been determined in the early 1930s—some forty-five years after the Dutch researcher Christiaan Eijkman began his work on Indonesian chickens—by Robert R. Williams, an American chemist at Bell Telephone Laboratories (and the brother of Chandler Medal recipient Roger Williams). Robert Williams had been investigating the vitamin's structure in his spare time for years, at one point studying it in his garage, using his wife's washing machine as a centrifuge. Williams christened the substance "thiamin" (*theion* is Greek for sulphur, which thiamin contains) and led the team that first successfully synthesized it in 1936. In 1939, a crystallized form of B1 was developed that could be added to food.

Today, as we saw earlier, the precise details of how thiamin works in the body are still not completely understood. But it's safe to say that we've moved beyond Vice President Wallace's description of its function, which he summarized to the conference like this: "It is the oomph vitamin!"

Wallace's enthusiasm might not have done much to counter his zany reputation, but at the time, relating food to oomph was not that odd—and America's thiamin supply in particular was thought to be a legitimate concern. The *Journal of the American Medical Association* had published a report in 1940 that estimated that sugar and white bread had accounted for 50 percent of Americans' calories in the preceding years, and health experts were worried that Americans' increasing taste for these vitamin-poor foods was putting them at risk of deficiencies in thiamin, since modern milling techniques were removing most of the B vitamins from grain (just as they still do today).

This was the same concern that Elmer McCollum had been raising since the 1920s and which had been embraced by the yeast cake campaign, but the most vocal of the thiamin pushers was now Russell Wilder, the physician who thought that the Nazis were withdrawing vitamins to crush conquered people's spirits. Wilder was no quack: in addition to being head of the Department of Medicine at the Mayo Foundation, he organized and chaired the Food and Nutrition Board of the National Research Council and was involved in some of the first clinical investigations of insulin; by the end of his career, the University of Chicago, the American Diabetes Association, and the American Medical Association had all given him awards.

But on the subject of vitamins, Wilder's views were extreme. He was convinced that thiamin could affect morale (as well as oomph, pep, and zest) and that Americans' supposed thiamin deficiencies were the country's greatest nutritional wartime threat. According to Wilder, the only way to conquer this deficiency was by adding thiamin back into people's diets, ideally via enriched flour. (As a reminder, enrichment generally means replacing micronutrients that processing has destroyed; fortification means adding micronutrients at higher amounts than were originally present or introducing micronutrients to foods that never naturally contained them.) Wilder pushed this idea to any audience that would listen, from reporters at the *New York Times* to a conference of the American Gastroenterological Association in Atlantic City. While at first he was opposed to adding thiamin to foods that didn't originally contain it, eventually he even supported the idea of fortifying fruits and vegetables.

Now, it's true that micronutrient deficiencies can affect a nation's health, weakening people's resistance to infectious diseases, stunting children's growth, and occasionally leading to unexpected and disastrous results, as when unfortified margarine contributed to an outbreak of night blindness in Denmark in World War I. And, as noted, the vitamin destruction caused by processing has indeed lowered the nutritional quality of our food supply.

But just as Elmer McCollum often assigned powers to vitamins that went far beyond what had been observed in humans, Wilder's beliefs in thiamin's abilities went way beyond reality, making thiamin a particularly striking historical example of the types of exaggerations that are so prevalent today and how willing we are to accept them.

The press in particular seized on the idea of thiamin as a "morale" vitamin that was "essential for growth, good nerves and youthfulness" and "steadie[d] your nerves, g[ave] you energy and restore[d] your zest for living if you've lost it!" It provided "charm, composure and good digestion," said the *New York Times*, while still managing to "stimulate without a letdown" and be "vitalizing—and supposedly beautifying" as well. "[A]long with national unity and national faith in democracy, there is another most potent morale booster," reported the *Times* in 1941, "the name of which is Vitamin B1."

Columns in the *New York Times* celebrated the arrival of thiamin-spiked malted milk, chocolate syrup, and peanut butter, and offered readers thiamin-rich recipes with names like "Muffins for Energy" and "Liver for Health." "No Thiamin, no pep. Instead, fatigue, nervousness, often other handicaps," said a 1940 ad for Quaker Oats. When a study (in which Wilder was not involved) found that people who ate thiamin-rich diets could hold their arms out from their sides longer than those whose diets were thiamin-poor, it inspired sports teams like the New York Rangers and the St. Louis Cardinals to give their players what the Cardinals' general manager called the "Wham" vitamin. The National Doughnut Corporation approached the Nutrition Division of the War Food Administration with a proposal for thiamin-fortified Vitamin Donuts, with an ad showing cherubic school children gazing longingly at the treats like sugar-crazed angels. It aimed to emphasize "the great morale value of donuts." In 1940, a man named Andrew Viscardi successfully filed a patent for thiamin-enhanced tobacco.[3]

Wilder also succeeded in convincing the government, and thanks in large part to his influence, politicians began to push millers to enrich their flour. In 1941, the effort succeeded: most of the millers—worried that they'd be accused of weakening the country's defenses—agreed to start producing bread flour enriched not just with thiamin, but also iron, riboflavin, and nicotinic acid, which they successfully fought to have renamed "niacin" to avoid the confusion with nicotine. But there was still plenty of flour *not* used specifically for making bread, and Wilder wanted that to be enriched with thiamin too. He succeeded again: by 1942, almost all America's bread was being made with enriched flour. Though it's no longer mandatory, most of America's flour is still enriched today.

By the time I heard about the vitamin doughnuts, I was curious about what had inspired Wilder's initial obsession with thiamin—whatever it was must have been pretty convincing, I thought, considering how passionately he

---

[3] Good news for nutritionally starved doughnut and cigarette lovers: a company called VitaCig has launched a line of vitamin-enhanced electronic cigarettes. And RSuper Foods, owned by former Pittsburgh Steeler and Pro Football Hall of Fame running back Franco Harris, has created the Super Donut ("#1 in Nutrition") fortified with numerous vitamins and minerals, including 30 percent of the RDA for thiamin. (The company also offers cinnamon-flavoured Super Buns.)

thrust his weight behind it. And indeed, when I looked into it, it seemed that much of Wilder's enthusiasm (and the ensuing public and government support) rested on a base that in concept, at least, sounded convincing: science.

It appeared that much of Wilder's conviction was rooted in the results of two studies that he had helped to run. After five weeks on a low-thiamin diet, the subjects in the first study had supposedly developed symptoms that ranged from weight loss, lack of appetite, and fatigue to constipation and "inconstant tenderness of the muscles of the calves." In the second study, after six weeks of a low-thiamin diet, subjects were demonstrating symptoms including insomnia, vomiting, dizziness, and the "reawakening of psychotic trends." When two of the subjects were then given thiamin, they experienced "a feeling of unusual well-being associated with unusual stamina and enterprise." And when the subjects received one milligram of thiamin hydrochloride, improvement was "observable in every case within hours." The subjects got back to their normal activities, "and apathy was replaced by lively interest in ward work and current events."

The results of the first trial so inspired the editors of the *Journal of the American Medical Association* that they published an editorial in 1940 titled "Vitamins for War," which promoted the idea of enriching flour with thiamin. The states of mind and body observed in the study's subjects, it said, "were such as would be least desirable in a population facing invasion, when maintenance of stamina, determination and hope may mean defeat or successful resistance." This, the editorial concluded, "suggests that efficiency for prosecution of a war can be increased by the simple expedient of providing a very little more of vitamin B1 than the public is receiving." This editorial alone was enough to make advertisers swoon.

But just as science, when properly applied, can provide valuable support, the word "science" itself can also be a crutch, suggesting a legitimacy that may not actually be there. Wilder's trials took place before the age of randomized, blind, controlled studies (they were none of these things). And the glowing press reports—not to mention the *JAMA* editorial—left out some important details.

The reality is that Wilder's first trial only included four people, all of whom were young female patients at a Minnesota state mental hospital. What's more, as the study's authors acknowledged, the diet they were eating was "more deficient in vitamin B1 than commonly is reported in association with the syndrome of beriberi." With that level of deficiency—which, by the way, was not at all common in the typical American diet of the time—it makes sense that they might have experienced some deficiency symptoms.

As for the second study, its participants were six female mental patients who worked on the cleaning staff at the Mayo Clinic, where Wilder and the researchers were based. Their ability to "work" was measured by having them do chest presses. The women's mental illnesses were deemed "quiescent" enough for them to participate, and their diet was indeed low in thiamin. But it was also unbalanced in other ways that could have affected their mental states: it was based primarily on nutritionally poor foods like white flour, sugar, tapioca, cornstarch, and washed polished rice—and explicitly allowed the participants to eat as much candy as they wanted.

Wilder himself was not deterred by the studies' small sample sizes, potential biases, or the prior mental status of their subjects. Nor did journalists seem interested in the details of the studies—as historian Harvey Levenstein points out, no one revealed that Wilder's claims were "based on a study of the mental states of ten patients in what was then called an insane asylum." Instead, America had allowed itself to be swept into the thiamin craze.

Unfortunately for Russell Wilder, thiamin's trendiness was short-lived. Subsequent studies did not find any increase in oomph (or pep or zest, for that matter) from extra doses, few people turned out to actually be deficient in it, and if Americans during World War II were lacking anything, it certainly wasn't morale. Today, thiamin is usually lumped along with seven other vitamins into the unexciting-sounding B complex; indeed, now that beriberi is no longer a common problem, not many Americans think about it at all.

But though the idea of a thiamin frenzy might at first sound ridiculous, it's not as foreign a concept as it might seem. For while "vitamins are not 'just another food fad'"—as the *New York Times* pointed out in 1941 in reference to their genuine necessity for human health—independent vitamins themselves still move in and out of trendiness in a cycle of vitamins du jour. The hot vitamin of the early 1920s was vitamin A; in the late 1920s it was vitamin C, and in the 1930s it was riboflavin, otherwise known as B2. Recent decades have seen the focus shift from C to E to beta-carotene (which our bodies can turn into vitamin A) to D, and I recently received a prediction from a market research company that vitamin K is next. Indeed, if you swap the word "thiamin" with today's trendy vitamin (or dietary chemical), those early ads and product claims don't sound quite so strange at all. "Increase your Energy Level with Vitamin B12!" said a poster I recently saw hanging in the window of a medical testing facility. For $25, I could get an injection of B12 that it claimed would lead to better sleep, clearer skin, lower stress, and an accelerated metabolism—no yeast cakes required.

And food manufacturers today still take full advantage of vitamins' ability to cast a healthy aura over products that have little nutritional (or taste) value of their own. Take WhoNu cookies ("Now delicious is nutritious, too"), a vitamin-enhanced line of cardboard-tasting treats that claims that each three-cookie serving contains as much vitamin E as two cups of carrot juice, as much vitamin B12 as a cup of cottage cheese with fruit (note: fruit contains no vitamin B12), and, my favourite, "as much Fibre [*sic*] as compared to a packet of fortified oats cereal, prepared in a bowl with boiling water." It's tempting to make other, more accurate comparisons: as much saturated fat as a Reese's Peanut Butter Cup! As much flavour as a shingle!

The truly odd thing is not that this type of product still exists, but that we don't challenge its claims. We buy foods that claim their added vitamins and dietary chemicals will "support a healthy metabolism," without demanding an explanation of how that has been proved—or what it actually means. We feel reassured when a food or cosmetic has been studied in "clinical trials," even if the label provides no information about where or how said trials were conducted, or what they actually found. Instead, inspired by the same joint forces of hope and fear that Elmer McCollum, yeast cakes, thiamin, and war evoked in our predecessors, we continue to believe that if we follow the right experts' advice, we'll be able to stave off sickness and disease; as long as advertisements employ the magic word "science," we are willing to accept claims that otherwise might crack under the pressure of common sense.

While we're still susceptible to nutritional fads, what's particularly interesting about the thiamin craze is that it would not have been possible just several years before Wilder began his B1 push. That's because until 1934, when Hoffmann-La Roche launched Redoxon, the world's first large-scale synthetic vitamin C product, most vitamins in fortified foods and supplements were extracted from the natural sources mentioned earlier—vitamin C from rose hips, for example, and vitamin A from fish-liver oil. (Vitamin D also came from fish-liver oil, or could be produced by an irradiation process patented at the University of Wisconsin in the 1920s.) These extraction methods had quality and scalability issues common to many naturally derived products: production levels are limited by the availability of raw materials; the concentration of vitamins in foods—usually low to begin with—fluctuates both seasonally and between crops, and vitamin extracts are often highly perishable.

But by the eve of World War II, many of the vitamins had been successfully chemically synthesized and prices had already begun to drop, thus removing many of the variables limiting their use. The first multivitamin product appeared on the market in the mid-1930s, and despite the fact that then, as now, most nutritionists recommended getting vitamins from whole foods instead of supplements, the idea of nutritional protection in a pill proved too alluring for people to resist.

Pharmaceutical companies like Roche, Merck, and Pfizer were soon pumping out vitamins, and by 1938, vitamins and multivitamin products had become one of Roche's major sources of income. Several years later, the *Journal of the American Medical Association* wrote that the "vitamin gold rush of 1941" put that of 1849 to shame.

Just as we do today, consumers saw references to vitamins everywhere—in ads in magazines, in newspaper articles, in prominent displays at their local drugstore. Vitamins even crept into the workplace: their manufacturers persuaded employers in war industries like airplane factories to give vitamins to their employees and came up with alliterative products like "pep pills" and a "Vitamins for Victory" pack of three capsules that employers could provide for their workers for three cents a pop. One union included a demand for employer-provided vitamins in its contract negotiations. Even non-war-related employers experimented with supplying their workers with vitamin pills: the Columbia Broadcasting System (CBS), for example, paid for vitamin supplements for its entire workforce.

Despite the specificity of McCollum's articles, government educational attempts, and food advertisers' claims, the public's actual understanding of vitamins remained extremely vague—a 1941 Gallup poll found that 84 percent of homemakers couldn't tell the difference between vitamins and calories. But advertisers likely thought that was just fine: it implies that the women equated vitamins with energy.

And besides, if you can meet your nutritional needs—not to mention become a better person—just by taking a pill, then who really needs to understand the details? "[W]hile the vitamin pills are costly, each pill is believed to be a miracle of concentration," wrote Robert Yoder, a journalist who coined the term "vitamania" in 1942 to describe the trend. "When the customer takes one of those, he may not know what he is taking, but whatever it is, he believes he is taking 10,000 units of it, each unit representing the thyroid glands of an entire herd of rare Andalusian mountain goats. A jolt like that is a bargain at any price." The advent of synthetic vitamins, in other words, set the stage for our modern obsession with nutritional shortcuts, our desire to find magic bullets that obviate the hassle of changing what we eat.

As a result of all these factors—including the popular press, the government's public health campaigns, food manufacturers' advertisements, the war, the thiamin trend, and the general availability of vitamins in pill form—sales of vitamin supplements surged. Between 1931 and 1939, vitamin sales in the United States grew from $12 million to about $82.7 million a year. By 1942, vitamin sales in the United States were about $136 million a year and nearly a quarter of Americans were taking vitamin pills. By 1943, yearly sales had increased to $180 million. As a writer from the *New York Times Magazine* summarized, "A mild popular interest in vitamins which began twenty years ago has developed into a wave of nutritional reform."

One group, however, was notably wary about jumping on the synthetic-vitamin bandwagon: the manufacturers of processed foods. This may seem surprising, given that food manufacturers had originally celebrated the discovery of vitamins, rushing to point out any and all vitamins that their products might happen to naturally contain. As one Kellogg's ad campaign had put it, "Get your vitamins in food—it's the thriftier way." But the same period of time that saw the advent of synthetic vitamins also saw the advent of better methods to measure the vitamin content of foods. And from the food manufacturers' perspective, the results were not good.

Although American politicians' fears over widespread nutritional deficiencies were overblown, it was true that the growing popularity of processed and refined foods was negatively affecting the nutritional value of the country's food supply (in other words, Wilder wasn't totally nuts in his concern). Milling flour, for example, really does remove 70–80 percent of its natural thiamin content, and in 1940, refined wheat flour was responsible for approximately a quarter of the average American's caloric intake (average per-person consumption was two hundred pounds per year). New measurement techniques and standardized units of measure were revealing that many vitamins are sensitive to heat, light, temperature, moisture, and time, variables that often come into play in food processing; indeed, the purification and refinement that's necessary to ensure stability, sanitation, and a long shelf life often cannot be accomplished without vitamins' being removed or lost. As companies were alarmed to discover, the processing necessary to create shelf-stable, sanitary packaged foods could destroy the very natural vitamins on which the advertisements for those foods were based.

Yet despite the emerging consensus among scientists that food processing could remove and destroy vitamins, many processed food manufacturers initially stuck their heads in the sand, steadfastly denying that their products were in any way nutritionally inferior to their less refined peers. At first, you couldn't entirely blame them—with no synthetic vitamins available for fortification or enrichment, there was no way to replace the vitamins that processing had destroyed. The problem seemed unsolvable.

But even as it became technologically and financially feasible to add synthetic vitamins to processed foods, some food companies were still reluctant to do so, partly out of concern that the implicit marketing message of fortified foods—that the extra vitamins made them healthy—might also imply the opposite: that unfortified foods were not. The growing popularity of vitamin pills seemed to back up this fear, since if consumers didn't think there was something missing from their normal diets, why would they need a pill? This was such a worry that food companies and manufacturing associations had actually co-opted Elmer McCollum—who for years had been warning against the dangers of refined white bread—into becoming a spokesperson for their cause. By the early 1930s, the same man who had earlier stated that "there is no justification for the demand for white flour by the public" was being paid by the National Bakers Association to advise it on how to improve the reputation of white bread. In 1938, he accepted a $250,000 pledge (in 1938 dollars) from the Grocery Manufacturers Association for the creation of a nutrition foundation, one of whose goals was to fund research to prove that food processing didn't remove vitamins—an approach reminiscent of early tobacco company-sponsored research designed to show that cigarettes had no ill effects on health.

But as is quite obvious today, food manufacturers did eventually change their attitude. They had to. The truth about food processing's effects on vitamins was being scientifically proved and beginning to leak into the public press, and with World War II looming, no food manufacturer wanted to be accused of weakening America's defenses. As a top vitamin researcher advised flour millers and sugar refiners, "to blink at the scientific facts, which will presently become common knowledge, will be suicidal for the commercial enterprises involved." Indeed, in 1939 even the American Medical Association urged manufacturers to "restore" the nutrients of processed foods.

And so, using both extracts from natural sources and synthetic vitamins created by newly invented chemical techniques, food manufacturers began to jump onto the vitamin bandwagon, adding them to their products and widely publicizing their addition. Whereas a 1933 ad for Cocomalt only boasts of its being a "rich source of Sunshine Vitamin D," for example, by 1940, ads claimed it could provide vitamins D, A, and B—at a recommended dose of three glasses per day.[4] Perhaps this about-face was a recognition of the public relations time bomb that could explode if and when the public learned the truth about vitamins' vulnerabilities during processing; perhaps it was simply an attempt to capitalize on the trendiness of the things. Regardless, one thing is clear: once food companies began to fortify and enrich products—and to recognize the enormous profits that vitamins could help them achieve—they never looked back.

In retrospect, food companies' embrace of vitamins occurred at a very fortuitous moment. Like a repentant thief replacing stolen silver before anyone noticed it was gone, manufacturers could replace the vitamins that food processing had removed before most of the public knew they were missing. Rather than being perceived as correcting a flaw, fortification and enrichment could therefore be advertised in purely positive terms, the addition of something good rather than the correction of something bad. As a result, food

---

[4] This also led to a new twist on food manufacturers' attempts to capitalize on nutritionists' recommendations that vitamins were best acquired from foods. An 1940 Cocomalt ad titled "Tommy Needs Vitamins—And *I* Need an Adding Machine!" for example, stresses that Tommy should get his vitamins "in *food form*, because many biologists say they're better assimilated that way"—conveniently not pointing out that many of the vitamins in Tommy's three daily glasses of Cocomalt were synthetic additions.

companies changed their tune toward vitamin measurement techniques as well. Now they were seen as tools rather than liabilities, able to confirm marketing claims of what nutrients food processing had added instead of revealing what it had destroyed.

The serendipity of this timing still affects us, for a reason that is both straightforward and profound: processed-food manufacturers have never had to publicly admit the potential nutritional inferiority of their products. Instead, they were—and are—free to highlight the benefits of enrichment and fortification without acknowledging why the addition of vitamins (or minerals) was necessary to begin with—even arguing, as the president of the Grocery Manufacturers Association did in 1956, that "today's processed foods have a food value at least equal, and often superior to, raw produce." And indeed, fortified and enriched products are still largely perceived in one of two favourable lights: they're just as good as their unprocessed, unrefined counterparts, or they're even better. The flip side—that this fortification is necessary because refinement has made the food inferior—is rarely acknowledged or discussed.

6

Nutritional Blindness

[A]t least two billion [people] experience "hidden hunger"—
they might get enough calories, but they do not get enough of
the vitamins and minerals their bodies need. The signs of
vitamin deficiencies . . . include birth defects, anemia,
blindness, impaired physical and mental growth,
maternal and child death, brittle bones and
increased susceptibility to disease. These
affect not only health but an individual's
future potential.
—KLAUS KRAEMER, DIRECTOR OF SIGHT AND LIFE, 2012

While we are often blind to the man-made vitamins in our own food
supply, there are many places in the developing world where the
absence of synthetic vitamins is all too evident. The story of one
vitamin in particular raises questions both about why vitamin deficiencies
still devastate so many lives today, and about how those deficiencies could
and should be corrected.

Visit a village in many areas of sub-Saharan Africa or South Asia at twi-
light, and you may notice something strange: groups of children who have
played together all afternoon will bifurcate at dusk. Some will continue their
games, running around in the fading light, while others will retreat to their
families' huts, sitting with their backs to the corner. They won't reach for
nearby toys; they won't even move for food. Instead they will remain in place,
eyes blankly staring, until a friend or family member brings food to them or
guides them away by the hand.

The reason for their stillness is simple: they can't see. These children are
suffering from what's called night blindness. As darkness falls, their vision
fades to black, leaving them vulnerable and disoriented. Local women in late
stages of pregnancy often suffer from it, too, making it difficult for them to
gather firewood after dark or prepare meals for their families. In some

villages, night blindness is so common in the third trimester that it's considered a normal part of pregnancy.

Night blindness is the first stage of xerophthalmia (from the Greek word for "dry eye"), a condition also known—in more pronounceable form—as nutritional blindness. At first, night blindness will only occur after days spent in bright sunlight. Then it will occur after sundown every day, regardless of how overcast the weather is. Left untreated, it will eventually be accompanied by dry eye, a condition in which the eyes can't produce enough of the lubricating mucus that protects them and keeps them moist; as a result, the eye's surface becomes keratinized, like skin. With no mucus, and with these skinlike cells on the surface, the cornea—which is the transparent coating of the front of the eye—becomes rough and dry, and develops worn spots called corneal ulcers. If still not treated, these ulcers can penetrate the eye, causing a hole through which the inside of the eye leaks out. Or, even more severely, the cornea may melt away entirely, destroying the eye and causing permanent blindness. This final stage can take less than a day.

Night blindness would have been familiar to scurvy-afflicted mariners, for it occurred on their voyages as well—and while it took longer to develop than scurvy, it was similarly terrifying. "The nocturnal blindness is at first partial, the patient is enabled to see objects a short time after sunset, and perhaps will be able to see a little by clear moonlight" wrote a French professor of medicine in 1856. "At this period of the complaint he is capable of seeing distinctly by bright candle-light. The nocturnal sight, however, becomes daily more impaired and imperfect, and after a few days the patient is unable to discriminate the largest objects after sunset or by moonlight; he gropes his way like a blind man, stumbles against any person or thing placed in his footsteps, and finally, after a longer lapse of time, he cannot perceive any object distinctly, by the brightest candle light."

No one knew the cause of the sailors' blindness (homesickness? humidity? masturbation?), and treatments were equally haphazard: they included putting caustic substances around the eyes to cause blistering (the irritation was supposed to prompt healing), applying a solution of silver nitrate to the tip of the penis (to stop "the improper use of the genitals"), and locking sailors for days or weeks into a *cabinet ténébreux*—a dark closet—with hopes that the eyes were simply worn out from bright light and would recover after being given a rest. This last treatment did occasionally prove effective, though one wonders how many "recoveries" were simply a desire to get out of the *cabinet*.

True cures were also discovered: the ancient Egyptians successfully reversed cases of night blindness with raw liver; in the late nineteenth century, scientists recognized it could be cured with cod-liver oil. But today there's a much more palatable treatment. It's so powerful that just one dose, squeezed from a gelatine capsule into a child's mouth, can cure night blindness in a single day. Its protection can last for half a year or more. It costs about two cents per capsule, and the FDA doesn't even consider it a drug. The miracle cure is vitamin A.

Vitamin A is a clear, fat-soluble molecule that's found naturally only in animal products, including organ meats, whole-fat milk, butter, and yes, cod-liver oil. Our bodies can also make it from a family of chemicals called carotenoids (particularly beta-carotene), which are pigments that give many fruits and vegetables their yellow, red, and orange colours. Vitamin A is stored in the liver, and most well-nourished people have enough to last up to a year. That's why it often took sailors longer to develop night blindness than scurvy: unlike vitamin A, the body doesn't have a significant reserve store of vitamin C.

Today, people in developed countries have easy access to all three sources of vitamin A: animal products like egg yolks and butter (plus beta-carotene-rich vegetables and fruits), supplements, and fortified foods like milk (since vitamin A is fat-soluble, most vitamin A in low-fat milk is a synthetic addition). As a result, severe vitamin A deficiency in the developed world is rare.

The rest of the world, unfortunately, is not so lucky. Vitamin A deficiency is an enormous problem in developing countries, and night blindness is a distressing indicator of its prevalence. Keith West, professor of infant and child nutrition at Johns Hopkins, estimates that one out of four preschool children in high-risk areas, primarily in sub-Saharan Africa and South Asia, are vitamin A deficient and at risk of going permanently blind. (He puts the number at nearly 130 million preschool children; other estimates are even higher.) In addition, he estimates that vitamin A deficiency-induced night blindness strikes about 6.2 million pregnant women per year.

The connection between vitamin A deficiency and night blindness is now well understood, and has to do with our rods, the cells in our retinas that allow us to see in low light. Rods contain a pigment called rhodopsin, also known as visual purple for its reddish-purple colour, that immediately bleaches when it's exposed to light—similar to what happens if you expose undeveloped photographic film to the sun. This bleaching breaks down the visual purple into new chemicals that translate light waves into nerve signals in the brain that create the images we see.

Our bodies need vitamin A to recycle these chemicals back into visual purple so they can be used again. You can actually observe this recycling happening whenever you're blinded by a bright light like a camera flash: the sudden light bleaches the visual purple in your rods, and it takes several moments for your body to "unbleach" it—using vitamin A—so that you can see again. Our bodies are able to recycle most of the vitamin A, but not all of it; every new unbleaching cycle uses up a bit more. If this vitamin A isn't replaced by your diet, you won't be able to replenish your rods' visual purple. And if you can't replenish your rods' visual purple, you won't be able to see in low light. You become night blind.[1] As Thomas Moore wrote in his 1957 book on the vitamin: "It may be an inspiring thought . . . that Man's knowledge of the existence of the stars and the vast universe which appears in the heavens each night, comes in the first place from the stimulation by the light rays of delicately poised molecules of vitamin A."

Vitamin A's role in preventing other symptoms of full-fledged nutritional blindness like dry eye and corneal ulcers is less well understood, but it likely has to do with the vitamin's role in maintaining our body's mucosal epithelial linings, the layers of cells that surround and protect our organs—including, importantly, our eyes. Vitamin A is so important for these epithelial linings (and works so fast) that if you catch a deficient person with corneal ulcers before his corneas have melted, a high dose of vitamin A can heal his ulcers within several days. It's so effective, so simple, that it's nearly miraculous. But it turns out that preventing and curing nutritional blindness is not the most amazing thing vitamin A can do.

Not so long ago, I found myself sitting across a table from the man responsible for recognizing some of the other powers of vitamin A—which offer further

---

[1] Vitamin A's role in vision has led to the folk wisdom claim that carrots help you see at night. This is partially true, but the reality is more complicated: the beta-carotene in carrots can be converted to vitamin A, which is essential for night vision, but unless you're starting off deficient in vitamin A, extra carrots aren't going to make a difference in your night vision.

proof of just how devastating vitamin deficiencies can and continue to be. Born in Brooklyn in 1942, Alfred Sommer is a renowned ophthalmologist and epidemiologist at the Johns Hopkins Bloomberg School of Public Health, who still maintains a hint of a New York accent even though his career has carried him all over the world; behind his desk is a collage of artistic black-and-white photos he took in his early travels, from India to Afghanistan to Tanzania to Tibet. There are awards, too—a shelf and wall of them, including the emblematic *Winged Victory* trophy of the Lasker Award, one of the most prestigious prizes in medical science.

Sommer's road to the Lasker could be said to have begun in 1976, when Sommer—whose previous daring adventures had included studying cholera and smallpox outbreaks in what's now Bangladesh (during a devastating cyclone and a nine-month civil war)—moved to the Indonesian city of Bandung with his wife and their five-month-old baby. His goal was to answer some of the enduring questions about nutritional blindness.

"We didn't know how common it was, we didn't know whether giving a dose of vitamin A twice a year would actually prevent it, and we didn't know why some kids were vitamin A deficient and others were not," he told me as we sat together in his office in Baltimore. So he did what any highly ambitious, energetic, and slightly obsessive ophthalmologist would do: he designed and launched three large-scale studies in Indonesia that he hoped would answer these questions.

This was not the first time he'd grappled with vitamin A. The World Health Organization, which is the agency of the United Nations that's focused on international public health, puts together nonbinding, evidence-based policy recommendations for its member states, including suggested treatments for noncommunicable diseases like vitamin deficiencies. Several years before he arrived in Indonesia, Sommer had served on a panel that had updated the WHO's treatment recommendations for nutritional blindness. With his input, the committee had changed the previous treatment recommendation for nutritional blindness—intramuscular injections of oily vitamin A, which "sat there like a lump and didn't do anything"—to injections of water-based vitamin A that the body (which, after all, is mostly water) can actually absorb. So when Sommer encountered his first seriously deficient patient in Indonesia, a child with corneal ulceration who was at immediate risk of going blind, he followed these guidelines and asked for a vial of water-based vitamin A.

"Here I am in Indonesia, freshly arrived, and here's the first kid with a corneal ulcer—that's an emergency—and everyone looks at me like I'm crazy," he remembered. "It turned out that there was no water-miscible vitamin A." Not just in the hospital where he was. Not just in Indonesia. Despite the fact that the WHO recommendations had been in place for several years, water-soluble vitamin A wasn't being produced commercially anywhere in the world.

Instead, the Indonesian hospital—and apparently everyone else—only had vitamin A in oily form. So Sommer, knowing that his patient would go blind if he received the oil as an injection, did something brash: he drew up a shot's worth of vitamin A and simply squirted it into the child's mouth. Then he called his contacts at the pharmaceutical company Roche and demanded that they come up with a water-soluble form of the vitamin—fast.

Within three months, Roche had succeeded in developing the new water-miscible form of the vitamin and was ready to distribute it to Sommer and other health workers. But in the meantime, Sommer had continued treating his patients with these oral squirts of vitamin A. To his relief and excitement, this appeared to work just as well as the water-miscible injections. So he designed a new study: he divided his patients into two groups, and treated one with oral vitamin A and the other with the new water-miscible shots. The response from both groups was exactly the same. ("It doesn't usually work this well in medicine," he said, laughing.) It was Sommer's first major vitamin A discovery, and he knew its potential impact was enormous: treating vitamin A deficiency by squirting a capsule into someone's mouth doesn't require a specially trained health worker or even a sterile needle. It's also far cheaper than injections: each capsule of vitamin A—enough to sustain a child for six months—cost only about two cents.

To Sommer, the conclusion was obvious: the WHO should update its recommendations based on his results, and say that vitamin A should be delivered by mouth rather than injection. But when he presented his data at a WHO/ UNICEF meeting in 1978, the assembled vitamin A experts disagreed.

"Their first argument was, 'Well, the kids could spit it out of their mouths,'" he told me. "I said, guess what, they *don't* spit it out because it's oily and it *sticks to their mouth and not a single kid didn't respond.*" (It was at this point in the conversation that Sommer's inner New Yorker began to more forcefully emerge.)

"Then the next argument was, 'Parents like an injection.' I said, look, fine, give them an injection, too. The problem is that no one's going to have water-miscible vitamin A on hand because it's expensive. What they're going to have is what they have now: oily vitamin A. And if the recommendations say the kids need an injection, then they're going to give them an oily injection and think they've done the job and *the kid will go blind*." It was a situation that's still common in medicine today: parental desires can negatively influence doctors' treatment decisions (such as when people demand antibiotics for viral infections, which antibiotics can't cure), and the medical establishment can be slow to change treatment recommendations based on the latest science.

The result? With "a lot of screaming on my part, they included a footnote in tiny letters saying that if there is no water-miscible vitamin A, you can use oily vitamin A by mouth," Sommer said, still frustrated by the backward emphasis. "It took ten years, *ten years*, to convince people to make the oral treatment the official recommendation, and put the injection in a footnote."

In the meantime, Sommer returned to the United States to join the faculty at Johns Hopkins, and was invited to become a member of the prestigious American Ophthalmological Society. But the invitation came with one substantial requirement: within three years from the upcoming February (it was October), he would be required to write a thesis and present it to the group. No thesis, and his membership would be denied.

Sommer decided that he'd finish his thesis before the clock even started ticking, which is how he found himself sitting in his office the week after Christmas in 1982, poring over green-and-white-striped paper printouts that covered an entire desk. They contained data from one of his Indonesian studies, an eighteen-month investigation following about 3,500 kids. Sommer was searching for some aspect of the data that could serve as the focus of his thesis, and as he stared at the numbers, he noticed something strange.

Typically, at each three-month follow-up, about 90 percent of kids showed up for their health examination; the ones who were missing were usually busy with chores or working in the fields. But Sommer noticed that at each successive three-month follow-up, children with mild forms of nutritional blindness (night blindness or dry eyes) were disappearing from his records in much more dramatic numbers than the kids with healthy eyes. Where had these children gone? Sommer looked deeper into the data. Eventually, by cross-analyzing different data sets, he realized something shocking: the kids with eye problems weren't missing because they were at the market or planting crops. They were missing because they had died.

The numbers were amazing. Children with night blindness were dying at three times the rate of those with normal sight. Those with Bitot's spots—dry spots on the cornea that are associated with nutritional blindness—were dying at six times the rate of the other kids. Children with both night blindness and Bitot's spots were dying at an astounding *nine times* the rate of their less deficient peers. It was what Sommer calls his "holy expletive deleted" moment: for reasons that couldn't be explained by blindness, the more vitamin A deficient the children had been, the higher their risk of death.

These results were particularly mind-boggling because at the time of Sommer's revelation, night blindness was thought to be an *early* sign of vitamin A deficiency—preventable and potentially dangerous but not life-threatening. (That's why it was considered ethical to run trials where children with night blindness were not all immediately treated.) Now it seemed not only that night blindness was a *late* stage of vitamin A deficiency, but also that children's vitamin A status had some important relationship with the strength of their immune systems and, thus, their resistance to infection. The lower their vitamin A levels, the higher their risk of developing severe, life-threatening infections like measles—and the higher the chance that those infections would kill them. Somehow, in other words, vitamin A seemed to be playing a role in the immune system that was keeping the children alive.

Sommer presented these findings at the ophthalmological society's meeting (presumably to his audience's surprise—"How often do you have an ophthalmological paper about child mortality?" he asked) and also published them in the *Lancet* in 1983. Though he couldn't yet prove that vitamin A deficiency had *caused* the children's deaths—the trial wasn't set up that way—the association he'd found was so provocative that it seemed to beg for follow-up studies. Instead, the opposite happened: it was so provocative that no one paid attention to it. There were no editorials, no studies. The letters to the editor in response to his paper totalled exactly one.

Frustrated, Sommer decided to try to attract more attention by producing more data. He was already in the midst of planning a randomized clinical trial of nearly 26,000 Indonesian children to prove that a twice-yearly large dose of vitamin A could reduce the risk of nutritional blindness. Now, he and his colleagues added death as one of their endpoints. When they finished the study, the results were even more astounding than he'd anticipated: the children who had received the vitamin A supplements had a 34 percent lower risk of dying than those who had not. The implication was dramatic. Provide children with adequate vitamin A—which his previous research had shown could be done just by squeezing a capsule into their mouths—and you could save not just their vision but their lives.

The *Lancet* published these findings in 1986, with an accompanying editorial of support. This time the medical and nutritional communities paid attention—and they weren't pleased. Angry letters to the editor poured in criticizing both Sommer and his studies. High doses of vitamin A can be dangerous, they said. He hadn't used a placebo control. And, most of all, his conclusion—that providing sufficient vitamin A might prevent 34 percent of deaths in deficient kids—was just too good to be true.

But here's the strange thing: it wasn't too good to be true. Nor was it a new idea. In fact, previous researchers had already discovered a connection between immunity against infections and what we now know as vitamin A. In 1904, a Japanese physician, Masamichi Mori, had noticed an association between dry eye and infections, and Danish pediatrician Carl E. Bloch had observed similar phenomena among Danish children in World War I. Biochemists including Thomas Osborne and Lafayette Mendel, two of the people who actually discovered the vitamin, had investigated a relationship between vitamin A and infection in rats. The pharmacologist Edward Mellanby—who is best known for his work on vitamin D and rickets—wrote in the *British Medical Journal* in 1928 that "it is, in fact, difficult to avoid the conclusion that an important, and probably the chief, function of vitamin A from a practical standpoint is as an anti-infective agent. . . . An extensive experience of nutritional work suggests that vitamin A is more directly related to resistance to infection than any other food factor of which we are aware."

Inspired by these other researchers' work, a British pediatrician named Joseph Bramhall Ellison launched a study in 1932 at the Grove Fever Hospital on children's measles, a highly contagious and often severe infectious disease that Ellison calculated was responsible for 50 percent of his patients' deaths. In what was probably the best study of vitamin A at the time, Ellison divided six hundred children with measles into two groups—one received the standard hospital diet; the other the hospital diet plus supplemental vitamin A—and followed them for the course of their hospital stays. The result? Vitamin A treatment appeared to reduce the children's deaths from measles by 58 percent.

As a result, vitamin A became known in the popular press as the "anti-infective" vitamin—a substance that Elmer McCollum described as being necessary to build fences that keep germs out. By the 1930s, the importance of vitamin A had become so recognized by the public that England's annual consumption of cod-liver oil (which is usually dosed by the teaspoon) had reached 500,000 gallons; the United States produced and imported 640,000 gallons of cod-liver oil in 1929 alone. In the minds of both nutritionists and the public, the connection between vitamin A and immunity was clear.

And then . . . it was forgotten. Not only had Alfred Sommer never heard of these previous studies when he began his Indonesian research—which is not entirely surprising, given that he was an ophthalmologist—but nutritionists had forgotten about them, too. How could this have happened?

Part of the collective amnesia might have stemmed from the fact that many early trials—there were more than thirty before 1940—had recruited people who weren't vitamin A deficient, and/or who had diseases that vitamin A can't help. Therefore, their results were inconsistent.

"Extra vitamin A doesn't do anything if you already have enough of it," Sommer explained. "But they were giving vitamin A like you'd give antibiotics, thinking it'd affect everyone the same no matter how much vitamin A they already had." (This is an important point: vitamin A will only help your immune system if you are extremely deficient in vitamin A, which very few Americans are, and it is toxic at high doses. *Do not start taking high doses of vitamin A.*)

Sommer's comment also touched on the other likely cause of the mass amnesia surrounding vitamin A and immunity: antibiotic drugs. The first antibiotics were developed in the 1930s and 1940s, and unlike vitamin A, which only produces marvels when a patient is already deficient in it, antibiotics' effects against bacterial infections were immediate, consistent, and astounding. The possible role of a vitamin as a preventive agent paled in comparison with antibiotics' curative powers.

But whatever the cause for nutritionists' forgetfulness, it wasn't until Alfred Sommer had his expletive-filled revelation—followed by expletive-filled years trying to get other scientists and nutritionists to believe his results—that the connection between vitamin A and immunity was fully recognized. Today, thirty-something years later, it's been more than accepted; it's been embraced. Vitamin A supplementation programs exist in more than seventy countries, and UNICEF estimates that these programs are reducing child mortality from all causes by up to 23 percent. In 1993, the World Bank called vitamin A supplementation one of the most cost-effective interventions in modern medicine, and in 2008 the Copenhagen Consensus, a group of notable economists who try to figure out cost-effective solutions to the world's most pressing problems, chose vitamin A (and zinc) supplementation as the number one investment the world could make to improve the state of the planet.

Yet despite overwhelming proof of its importance, despite all the progress that has been made, we still don't entirely understand the intricacies of how vitamin A works in our bodies. It appears to regulate at least several hundred genes, which likely accounts for its importance in the proper growth and differentiation of tissues throughout the body, especially the mucous epithelium, which lines the outside of the eye and the respiratory tract (where it protects against infections, among other roles). Some of the genes that it regulates may control the creation of immune cells themselves. But whatever their details, these regulatory roles are completely different from vitamin A's function as a building block of visual purple. As is true for every vitamin—and nutrition as a whole—there's much we still don't understand. In the meantime, the WHO estimates that between 250,000 and 500,000 vitamin A-deficient children are still going blind each year. And within a year of losing their sight, says Sommer, up to 90 percent of those children may die.

So why are so many people in the world deficient in vitamin A? Sure, people in the developing world often do not have the same access to supplements and fortified foods as those in the developed world. But many of the people suffering from nutritional blindness aren't starving—they have adequate calories. And if official data about the availability of vitamin A in those people's foods is to be believed, then vitamin A deficiency doesn't seem like it should be such an issue. This discrepancy nagged at Alfred Sommer.

"What always bugged me was why vitamin A deficiency was so universal among kids in the developing world," Sommer told me. "If you look at the Food and Agricultural Organization [FAO] and the World Health Organization's food charts that tell you how much of each vitamin is in the food supply in a region, there's more than enough in Africa, and just about enough in Asia. And yet any time anyone did a rigorously controlled study, about half the kids would turn out to be seriously deficient, and there'd be about a third reduction in mortality [when you provided supplemental vitamin A]. Why was that?"

The answer turned out to be directly related to plants. The standing viewpoint of the time—and indeed, an idea that is still echoed today—is that kids in the developing world could avoid vitamin A deficiency by eating more vegetables and leafy greens that contained beta-carotene, which our bodies can convert into vitamin A. (If you look at a diagram of beta-carotene's molecular structure, it resembles two vitamin A molecules joined at their tails.) The calculations of the food supply charts were based on this conversion: if people had access to beta-carotene-rich vegetables, they presumably had access to vitamin A.

But reality seemed more complicated. Sommer remembers that in 1974, at the first vitamin A conference he attended, an Indonesian scientist presented research showing that feeding people an abundance of beta-carotene-rich dark leafy greens for three months didn't change their vitamin A status. The paper, the findings of which were considered heretical, was never published, but in the 1990s, a Dutch group basically replicated his work (as subsequent researchers have done as well). They fed large amounts of leafy green vegetables to women and children for three months—amounts that should presumably have provided a huge vitamin A boost because of their beta-carotene. Their levels didn't budge.

The problem, the researchers suspected, was the conversion. If you eat pure beta-carotene dissolved in oil (beta-carotene, like vitamin A, is fat soluble), the body requires only two to three beta-carotene molecules to produce one molecule of vitamin A. But if the beta-carotene occurs in a non-oily food—say, a part of a plant that doesn't contain much oil, like a leaf—the conversion is far less efficient. The WHO tries to take this discrepancy into account by using a conversion factor of 6 molecules of beta-carotene to 1 molecule of vitamin A. In the United States, the Institute of Medicine has settled on a conversion rate of 12–1. But in reality, the amount of beta-carotene we convert to vitamin A from most plants is far lower.

"What the researchers found is that if you feed someone a fruit that's rich in beta-carotene—papaya, mango—it takes about 18 molecules, not 6, to get one molecule of vitamin A," said Sommer. "And if it's a dark leafy vegetable like spinach, it takes about 27 molecules of beta-carotene to make one of vitamin A. So if you put together what a kid might get from fruits and leafy vegetables, it'll take about 24 molecules of beta-carotene to make one molecule of vitamin A." Other recent papers have come up with slightly different conversion factors, but the basic range is the same.

That answered Sommer's burning question: the WHO and FAO calculations were based on conversion factors two to four times higher than what appears to actually be happening in real human bodies. Run the numbers with an updated conversion factor, says Sommer, and you find that "in no place outside of modern economies do people get an adequate supply of vitamin A."

Providing vitamin A in the form of supplements or fortified foods is one popular solution to this problem—and indeed, supplements and fortified foods are largely responsible for the relative lack of vitamin A deficiency in the developed world as well (especially now that fried liver has fallen out of favour as a dinner option). Sommer has worked with governments and non-governmental organizations to organize and run supplementation programs that distribute twice-a-year vitamin A capsules to children, women, and other vulnerable populations; companies like DSM and its philanthropic nutritional think tank, Sight and Life, also provide fortified products, like vitamin A-enhanced sugar and MixMe packets of powdered micronutrients that people who are at risk of vitamin deficiencies can add to their food.

These programs do indeed save the sight and lives of many of the people who receive the supplements. But while these approaches are effective for the people they reach, it's impossible to provide supplementation to everyone who needs it—and since nutritional requirements are lifelong, those lucky enough to receive the supplements or fortified products once still need a regular supply. The ideal solution would be to provide a sustainable source of vitamin A (and other micronutrients) that didn't depend on regular outside interventions. But this, while it might seem logical, has also led to one of the most controversial subjects in nutrition.

One solution to the problem of vitamin deficiency in the developing world would be to boost plants' ability to make vitamins: if we could create vitamin-enhanced forms of popular crops, communities could grow their own micronutrients without reliance on external supplements or fortified food. Creating varieties of crops that produce higher levels of desired nutrients is more formally known as biofortification—fortifying nature.

In some cases, biofortification can be achieved through traditional breeding, also known as selection: you pick varieties of crops that have a desired characteristic—say, one sweet potato with naturally high levels of beta-carotene and another with good resistance to disease—and breed them till you end up with plants that express the characteristics you desire. Traditional breeding is responsible for nearly all the fruits and vegetables we eat today, producing everything from tasteless (but long-lasting and easily transportable) supermarket tomatoes to the wide variety of apples that arrive in farmers' markets each fall. Though it hasn't typically been a priority, it's often possible to use traditional breeding to improve the nutritional content of produce—the seed company Burpee, for example, recently launched a BOOST Collection, which is a selection of tomatoes, peppers, cucumbers, and lettuce that were chosen from existing stocks for their naturally higher content of vitamin C and several other plant compounds thought to be beneficial to human health.

But while traditional breeding is a powerful and relatively uncontroversial technique, it also comes with challenges: First, it's slow—it can take generations to create the micronutrient profile (or other characteristic) that you desire. Second, since you're introducing the entire genome of the selected plant, not just one particular characteristic, you may end up with undesirable characteristics in addition to the ones you want. For example, you could end up with a crop that has a higher level of a particular micronutrient but is also more vulnerable to disease. In that case, you might need to back-breed the plants to get rid of those undesirable traits—a process that can take years.

Perhaps most important, there are times when traditional breeding techniques simply won't work. Bananas, for example, are nearly impossible to breed because the ones we like to eat don't contain viable seeds. (Those little black dots in their flesh are undeveloped, defective seeds: they possess three chromosomes instead of two, which prevents them from fully forming.) Without seeds, a plant can't be conventionally bred—which means, obviously, that you can't introduce new traits through conventional breeding.

In other cases, the desired characteristic simply does not occur in nature—and you can't select or breed for something that doesn't exist. That's one of the problems with rice, which is a staple food for roughly half the world's population, providing the majority of calories eaten by people in countries such as China, India, Bangladesh, Cambodia, Laos, Myanmar, Thailand, Vietnam, and Indonesia—indeed, in those countries, many weaned babies eat little else. Unfortunately, as we've seen, white rice (which is the part of the rice kernel known as the endosperm) is remarkably devoid of most micronutrients—including beta-carotene.

Like other vitamins, beta-carotene helps plants absorb light and protects them from the oxidative damage caused by photosynthesis. Other parts of the rice plant that are involved in photosynthesis or energy production, like the leaves and the roots, *do* produce beta-carotene for these purposes. But the endosperm is buried deep within the rice kernel; as the plant's fuel tank, its purpose is solely to store energy—in the form of starch—for use by the germinating seed. From a rice plant's perspective, then, there's no reason for the rice endosperm to contain beta-carotene. This makes good sense for the plant, but from a human perspective, it creates a problem: anyone whose diet consists primarily of white rice is at risk of multiple micronutrient deficiencies, including not just beriberi but a life- and sight-threatening deficiency in vitamin A. What's more, since there are no varieties of rice plants that naturally produce beta-carotene in their endosperms, it's impossible to create beta-carotene-rich rice through traditional breeding.

But what if there were a way to develop a variety of rice plant that *did* produce beta-carotene in the endosperm as well as in its leaves? That became the challenge for Ingo Potrykus, now professor emeritus at the Institute for Plant Sciences of the Swiss Federal Institute of Technology, and Peter Beyer, a professor at the Center for Applied Biosciences at the University of Freiburg in Germany. Potrykus had been interested in the idea of beta-carotene-fortified rice since the late 1980s, motivated in part by his experience growing up in post-World War II Germany, where he and his brothers were often so hungry that they were left scrounging for food in the fields. In 1991, he started a doctoral program on the subject, and eventually met Beyer, who was an expert in beta-carotene production in daffodils (beta-carotene is what gives the flowers their cheerful yellow colour).

With the help of a grant from the Rockefeller Foundation, the two men began collaborating on the problem using recombinant DNA technology— what the world knows as genetic engineering—to try to develop rice that would produce beta-carotene in its endosperm. In 1999, after eight years of work and at Potrykus's mandatory retirement (in Switzerland, you must retire at sixty-five), they announced to the public that they and their research team had succeeded: by inserting four genes—one from a soil bacterium, two from daffodils, and one marker gene (that is, a gene used to indicate whether your other insertions have been successful)—they had created rice plants whose endosperms contained beta-carotene. The yellow-orange colour of the grains, which was caused by the beta-carotene, inspired its name: golden rice.

What was particularly surprising—and exciting—about golden rice's creation, said Potrykus and Beyer, was that they didn't have to insert a *complete* new metabolic pathway: the rice plants already had some of the metabolic pathway steps in place to make beta-carotene in the endosperm, which means that many of the genes required to do so were already being expressed. The trick was to figure out a way to complete the pathway. It was as if the scientists had begun to build a train track, only to discover, midconstruction, that the destination station was already there; all that was missing were a few crucial rails. When they reinserted those rails—in the form of the daffodil and soil bacterium genes—the rice began to produce beta-carotene in its endosperm in the same way it was doing in its leaves.

In 2000, the scientific details of golden rice were published in *Science* with a commentary that described it as "examplif[ying] the best that agricultural biochemistry has to offer," and *Time* magazine ran a cover story called "Grains of Hope" that celebrated the development of golden rice as "the first compelling example of a genetically engineered crop that may benefit not just the farmers who grow it but also the consumers who eat it." This was not long after Alfred Sommer's research had convinced the world that vitamin A deficiency was causing not just blindness but death—so figuring out a way to alleviate vitamin A deficiencies through biofortified crops was recognized as a potentially lifesaving idea for millions of people around the world. But as that same article pointed out, golden rice had no sooner been announced than it became known by another moniker: Fool's Gold.

One problem was that the amount of beta-carotene in the original version of golden rice was relatively low, which meant a very large amount of rice would need to be eaten in order to meet daily vitamin A requirements. But the bigger problem was that by that point, all genetically modified organisms (GMOs) had earned a reputation as Frankenfoods. Making things worse, Potrykus and Beyer had partnered with the biotech company now known as Syngenta to develop the rice. This was a necessary partnership since it helped them get legal access to dozens of processes that were vital for their research but whose patents would be violated if a product was to be developed. However, their mere involvement with a biotech company was enough to raise anti-GMO activists' hackles.

There are many legitimate questions to be asked about the general idea of transgenic—that is, GMO—crops. If you create a plant that is resistant to certain pests, for example, as is true for Bt cotton (cotton that's been engineered to be resistant to a common and devastating soil bacteria called *Bacillus thuringiensis,* and which now accounts for more than 90 percent of the US cotton crop), you may end up inadvertently breeding "super" pests that are resistant to the genetically modified crop and require stronger pesticides to control. Indeed, the widespread use of Roundup Ready cotton, designed to be resistant to the common pesticide Roundup (so that you can liberally apply pesticide without killing the cotton itself) appears to have encouraged the development of a new, potentially devastating generation of resistant weeds.

In addition, plants that have been engineered to manufacture their own pesticides and herbicides could turn out to be more hazardous to human health than their pesticide-laden counterparts—you can't wash off toxins if they've been incorporated into the plant. There are questions about potential allergens, and about genetically engineered crops' impact on local ecosystems, including what might happen if genetically modified seeds mingled with seeds from non-genetically modified plants. And there's also the issue of intellectual property: if farmers become dependent on seeds produced by one company, and if that company legally prohibits them from saving their seeds from one harvest to the next (thereby requiring them to buy new seeds every single year), then farmers could become permanently dependent on that one company to survive.

But as one of the original developers of golden rice, Salim Al-Babili, emphasized, it's impossible—or at least unwise—to make a lump judgment about *all* genetically engineered crops, and in his view, as well as that of many other scientists, golden rice has satisfying answers to most general GMO concerns. In the case of golden rice, he explained, scientists are not introducing a new product into the environment—beta-carotene exists in *all* green plants, including the rice plant itself. The inserted genes in golden rice are coaxing the plant into creating the same beta-carotene in its endosperm that it's already making in its roots and leaves.

Golden rice contains no added pesticidal or herbicidal properties. Nor is it likely to be toxic: beta-carotene is not allergenic, and while large doses of vitamin A can indeed be quite dangerous, our bodies are wise enough to know when to stop converting beta-carotene when vitamin A levels get too high.

As for the intellectual property issues, Syngenta and the other patent holders recognized the public blowback that would occur if they tried to directly profit from golden rice, and so had agreed that farmers with less than $10,000 in profit from the rice could use the golden varieties with no royalties or restrictions; they could save their seeds, and breed them freely with local varieties more suited to their climates and tastes. Perhaps the corporations hoped that positive press from golden rice might encourage public acceptance of some of their other engineered crops. But in reality, the opposite appears to have happened: public concern over the general topic of GMOs stalled field and human trials of golden rice.

The first non-greenhouse test field of golden rice was finally planted and harvested on a plot in Crowley, Louisiana, in 2004, after five years of regulatory hurdles. (Before that, researchers had to work in contained greenhouses.) In a photo taken around that time in the Louisiana field, Beyer and Potrykus look proud and hopeful. Not only had their field experiment been successful, but researchers at Syngenta had just produced a second experimental line of golden rice (GR2) that contained twenty-three times more beta-carotene than their original prototype—an improvement achieved by swapping the daffodil gene with one from corn. To Beyer and Potrykus, golden rice is a product born of pure philanthropy, meant not as a magic bullet but as a complement to other strategies to help solve the devastating problem of vitamin A deficiency.

But unfortunately for the people who might benefit from it, more than a decade later, golden rice is still not being grown outside research experiments. Instead, golden rice has been used by anti-GMO activists to fuel the overall controversy over genetically modified crops. Organizations like Greenpeace have mounted large public opinion campaigns against it, and there has been vandalism of test fields. Largely as a result of these smear campaigns, there have still been almost no studies of golden rice in people, which has made it impossible to do a true cost/benefit analysis.

Today, more than thirty years after Sommer noticed the connection between mortality and vitamin A deficiency, the truth is that golden rice should indeed be a controversial subject—but not for the reasons we usually think. What we should be asking is why we in the developed world feel entitled to pontificate on (and block) other people's access to potentially life-saving micronutrient interventions like golden rice, especially considering that so few of us are at risk of a vitamin A deficiency ourselves.

Perhaps the question answers itself—we are able to moralize about whether other people should be able to eat golden rice precisely *because* severe micronutrient deficiency is not a problem in our own lives. Indeed, our easy access to synthetic vitamins has caused us to develop a different form of nutritional blindness, one in which we don't recognize our own hypocrisy: we pass judgment on how other people should get their vitamins while we insist—especially when it comes to dietary supplements—that no one block access to our own.

7

From Pure Food to Pure Chaos

Not only must the consumer be not disfigured or killed,
he must get what he thought he was getting when he read the
label. As new standards for nutrition are set, as new truth
about food—some of it more fantastic than fiction—becomes
the substance for advertising claims, the consumer and
the honest business man may be assured that these
new truths will not be misused by charlatans.
—Paul McNutt, chairman, National Nutrition
Conference for Defense, "The Challenge of
Nutrition," May 26, 1941

The General Nutrition Corporation store near my childhood home—that's the official name of the nationwide supplement chain GNC—is one part science fiction and one part nineteenth-century apothecary. Its shelves are lined with products whose names are so hi-tech that they're incomprehensible, like vitaliKoR Daily Maintenance and Cellucor M5 Extreme. But despite the modern packaging, the fluorescent lighting, and the innumerable mentions of "science," there's something oddly anachronistic about it; when I visited it one blustery January afternoon, I felt like I was stepping back in time.

My purpose was dermatological: after three problem-free decades, I had developed sensitive skin. My hypoallergenic wedding ring gave me a rash, I had dry patches on my arms, and I had recently begun to suffer from intensely itchy calves. I'd tried seeking answers from modern Western medicine—I had done allergy tests for hundreds of chemicals and seen multiple dermatologists—but other than giving me steroid creams and coupons for Aveeno while billing hefty fees to my insurance company, no doctor had been able to help. And so I'd come to GNC with the same purpose that draws countless other people to its aisles: I wanted to find a solution in a supplement.

By supplement, I didn't necessarily mean a vitamin, though the two terms are often incorrectly used interchangeably. True, technically speaking, all vitamin pills are considered dietary supplements. But dietary supplements are not all vitamins—the term also includes nearly every other legal, nonpharmaceutical product that you could ingest for health, including herbs and botanicals, amino acids, enzymes, metabolites, and the ominous "organ tissues and glandulars" (which are exactly what they sound like: ground-up organ tissues and glands). Perhaps because of this definition's expansiveness, when I tell people I'm writing a book about vitamins, they rarely ask me about the ones found naturally in food. Instead, they equate vitamin with supplement, and assume that I'm talking about pills.

Even the signs labelling the aisles in many drugstores use the term "Vitamins" when what they really mean is "Dietary Supplements." But few supplement manufacturers care to clarify the distinction; on the contrary, vitamins' universally positive connotations mean that most companies are perfectly pleased to have their chondroitin supplements bask in vitamins' radiant glow. Consider the Vitamin Shoppe, one of GNC's competing chains. There are only thirteen human vitamins—and yet the so-called "Vitamin" Shoppe sells more than eighteen thousand products.

Despite their shared categorization, there are significant differences between traditional vitamin pills and the more exotic concoctions on GNC's shelves—whether they be Chinese herbs, botanicals, or proprietary products with crazy-sounding names. For example: We know how to identify vitamins chemically. We have a basic understanding of what they do in our bodies. The question of whether to take additional doses of vitamins as *pills* is controversial, but there's no question that the thirteen substances themselves are essential for human health. Their safety profiles have been studied, often in controlled trials; we have at least some sense of how they interact with drugs, as well as which ones, at approximately which doses, can make us sick. While they come in multiple formulations, there are no "proprietary blends" for standard vitamins; their ingredients are listed on the label, and nearly all of them have established RDAs (even if those recommendations are works in progress). Since most of the vitamins in supplements are produced synthetically, their potency doesn't depend on growing conditions, nor is their supply subject to seasonal variations (both of which can affect herbs and botanicals).

And at least in most cases, if a product says it contains vitamin C, it probably really does contain vitamin C.

All this is not necessarily true for non-vitamin and non-mineral supplements. And so, in the eyes of government regulators, I assumed they wouldn't be treated the same way.

I was staring at a homeopathic remedy display when a young salesclerk approached, wearing thick-framed glasses and a woolen hat.

"Can I help you find something?" she asked. I told her about my skin issues.

She thought for a moment. "Are you taking fish oil?"

Yes, I was.

"How much?"

I told her that I'd been taking a few capsules every other day or so, trying to get about a gram total of EPA and DHA, the two long-chain fatty acids in fish oil that have been associated with brain and heart health.

"Maybe you should try omega-7s, then," she said. "They're very important for the skin."

I looked at her blankly. In all my research on fatty acids—and as a health journalist with a long-term interest in fish oil, I have done a lot of research on fatty acids—I had never heard of omega-7.

"It's a different type of fatty acid—you can get it from sea buckthorn," she said.

"Sea buckthorn?"

"It's a plant that grows in really harsh conditions," she said, her tone suggesting that its ruggedness alone was a reason I should ingest it. She led me around the corner to a display of supplements and picked up two options, Supercritical Omega-7 and Sea Buckthorn Force. Since their ingredients appeared basically the same, my choice came down to product name: Did I want the judgmental sea buckthorns, or the ones that sounded like an elite military unit?

I went for the Supercritical.

"While the benefit of Sea Buckthorn on the body's largest organ, the skin, has been widely documented in ancient texts," said the box of the Supercritical formulation, "it also has an extensive history of traditional use in Tibet & Mongolia for the body's sensitive internal organs—in particular the mucous membranes that line the stomach/GI tract, upper respiratory tract, and vagina. The ability of this botanical to help moisturize and soothe these sensitive internal areas has made it a staple of traditional herbal systems. Similar properties have been reported in various modern studies."

Forget my sensitive internal areas—the description itself made me feel *emotionally* soothed. The dermatologists and allergists I'd seen had been annoyingly noncommittal, unable to even agree upon a diagnosis, let alone a treatment. In contrast, the box's label was decisive and reassuring, conjuring up images of ancient sages possessed of a wisdom that modern medicine has lost. And whereas modern medicine is full of caveats—"This is for some people but not others"; "It may be eczema but we can't be sure"—sea buckthorn was confident: it would solve my skin issues, no doubt about it. Let us not forget, after all, that its powers had been widely documented in ancient texts. It had an extensive history of traditional use in Tibet *and* Mongolia. It had been investigated in various modern studies. Like those of many supplements, the whole label evoked an aura of ancient Eastern wisdom just waiting to be rediscovered, a pharmacopeia of natural treatments ready to step in where Western medicine has failed. Despite my inherent skepticism, I couldn't help wondering, What if this stuff actually *worked*?

Merely posing the question to myself made me feel hopeful, a pleasant emotion that—at least in reference to my skin—I hadn't experienced in a long time. And as I looked at some of the store's other products—the milk thistle for liver health, the gingko for mental clarity, the Women's Procreation Vcaps ("nature's gift to hopeful moms")—it occurred to me that hope is the driving force behind nearly *all* of the supplements that line supermarket and pharmacy shelves. Viewed this way, the Vitamin Shoppe's eighteen thousand products begin to make a lot more sense: in a world where the so-called medical experts are often unable to help, we turn to supplements for comfort. The more desperate we are and the more modern medicine throws up its hands, the better "ancient wisdom"—and supplements that claim to encapsulate it—can begin to sound.

Unfortunately, when I read the sea buckthorn's box again, I realized something less inspiring: I had no idea what it was actually talking about. What was an "herbal system"? What did it mean to be supercritical? (Was I going to be super disappointed?) Since when has Mongolia been a top source of medical advice? And which "various modern studies" had investigated sea buckthorn's effects on vaginas?

I phrased my questions slightly differently for the salesclerk, asking her whether she knew of any controlled clinical studies involving sea buckthorn, whether there was a way to tell its quality, and whether it had been approved by the FDA. She didn't know about any studies, though she assured me that she was personally confident that the quality of that particular company's products was high. As for the FDA?

"They're more about pharmaceuticals," she told me. "When it comes to this type of stuff, they don't get involved."

The idea that the FDA wouldn't "get involved" in an industry that brought in more than an estimated $32 billion in US sales in 2012 seemed strange, to say the least. But the clerk was insistent—and in many ways correct. By the time I left the shop, sea buckthorn in hand, I had questions much larger than the mystery of my omega-7s. I wanted to know how we, as consumers, have become okay with the idea that vitamin C should exist in the same regulatory category as RIPPED FREAK Hybrid Fat Burner, or capsules filled with ground-up glands. And I wanted to know how we've reached a point, in this age of what can seem like ever-expanding government regulation, where dietary supplement makers in the United States—whose products go way beyond the thirteen human vitamins—are not required to do any testing for safety or efficacy at all.

The answers to these questions, while complicated, can be traced back to vitamins themselves—for while vitamins and supplements are not necessarily synonymous, the two categories are historically and philosophically linked. The discovery of vitamins introduced the idea that not only could natural substances found in food prevent and cure deadly diseases, but they could be downright *miraculous*, able to save lives with none of the toxic side effects of drugs. Thanks in large part to vitamins, the Hippocratic saying "Let food be thy medicine" took on an entirely new significance.

Later, the creation of synthetic vitamins both expanded the public's definition of "natural" (now nature could be found in not just foods and herbs but pills) and introduced the idea that natural food itself was inadequate and needed to be supplemented. Our desire to fill this supposed nutritional void—and our embrace of the ideas that food could be medicine, nature could come in pills, and anything natural must be safe—opened the door for the multibillion-dollar supplement industry that exists today.

This industry could not exist without a regulatory framework built to support it, and for that vitamins are partially responsible as well. Whereas the past century has seen America's drug and food regulations become progressively tightened, Americans' devotion to—and faith in—vitamins has given the dietary supplement industry a toehold to wildly expand the definition of dietary supplements while simultaneously loosening their regulatory requirements. The details of how this came to pass are surprising to begin with. But when you put America's supplement regulations in the context of our approach toward foods and (even more so) drugs, the story is even stranger.

The origins of America's supplement regulations could be said to date to 1862, before vitamins had been discovered or the idea of dietary supplements even existed. That's the year that President Abraham Lincoln created the US Bureau of Agriculture—and with it, the first Division of Chemistry, the organization that eventually evolved into the Food and Drug Administration, the agency better known as the FDA. (The FDA is the federal agency now responsible for regulating America's foods, supplements, and drugs, among many other things.) To say that the Division of Chemistry started small would be an understatement: its employees totalled exactly one—namely, a lone chemist who had previously been at the Patent Office testing fertilizers and animal feeds for adulteration and misbranding problems.

Despite this limited workforce, the need for a Division of Chemistry—and with it trained analytical chemists—was obvious from the start. By the mid-nineteenth century, adulteration in American commerce was spiralling out of control. Improved transportation and new technologies were shifting production from local shops and farms to distant factories, a depersonalization that made it easier for manufacturers to get away with cost-cutting substitutions such as diluting milk with water, mixing copper sulphate with flour (copper sulphate absorbs water, making bread seem heavier), or even adding ground-up lice to brown sugar (apparently the two have similar consistencies).[1] As adulterers increasingly turned to chemistry for new tricks, there was a growing need for analytical chemists who could catch them.

---

[1] Sometimes not all the lice were sufficiently ground up and killed, leading to a condition, caused when storekeepers scooped sugar containing live lice out of adulterated barrels, that was known as grocer's itch.

The gravity of the situation was made clear in a popular 1846 book called *Adulterations of Various Substances Used in Medicine and the Arts* that was meant as a how-to guide for pharmacists and physicians to test for adulteration in their products—and in the case of drugs, to see how they did or did not match substances listed in the *US Pharmacopeia*, the country's official record of accepted medications. The guide covered everything from food ingredients like cider, beer, cinnamon, coffee, milk, olive oil, and soda water to sealing wax and gunpowder. There were tips on ways to identify real arsenic ("when thrown upon hot coals . . . a strong smell of garlic is perceived") and whale spermaceti ("friable and somewhat unctuous"), not to mention Harry Potter-esque ingredients like Dragon's Blood, gall-nuts, croton oil, scammony, and asafetida—a fetid-smelling gum resin, the best kind of which was to be found "in large lumps."

As the most comprehensive resource of its kind, the guide was greatly useful to pharmacists and shopkeepers trying to confirm the quality of their stocks. But it did not solve the root of the problem: American products, especially medicines, were adulterated so frequently that they carried the same stigma that many Chinese products do today. European countries began embargoes against American food. At one point, adulteration was such a concern that the US Navy's Bureau of Medicine and Surgery stopped buying drugs from domestic pharmaceutical companies and began making them itself.

Adding to the problem of adulterated official drugs was the exploding number of "patent" or "proprietary" medicines on the market: untested, unregulated nostrums (short for Latin *nostrum remedium*—"our remedy") that any entrepreneurial American could bottle up for sale. Today, we often refer to these products as "snake oils" without realizing that snake oil was a real thing. Squeezed from actual snakes (though often diluted with other oils that were less dangerous to extract), it was sold as a patent medicine at the end of the nineteenth century for rheumatic pains, strains, sprains, and bruises, among other conditions.

Even without snakes, you'd think that names like Hamlin's Wizard Oil or Dr. Williams' Pink Pills for Pale People would themselves have given customers pause. But in the days when many diseases weren't understood or treatable, let alone curable, any medication that offered relief was likely to succeed—even if its relief was due to addictive ingredients like morphine and heroin, which manufacturers often included but weren't required to disclose.

In fact, manufacturers weren't required to disclose *any* of their ingredients, which meant that, despite their name, many of these "patent" medicines weren't patented at all. In the cases where manufacturers did bother to officially register their creations, they often focused more on the packaging than the contents, trademarking or copyrighting things such as the shape of the bottle, the box the medicine came in (both of which were crucial for illiterate customers), or the advertising copy used to sell it.

Patent drug manufacturers had plenty of time to design this packaging because there were no rules or regulations about the medications themselves. In the nineteenth century, there were no labelling requirements for foods or drugs. There were no quality-control guidelines or good manufacturing practices, no preapproval or safety testing. It would have been bad enough if the lack of oversight led to medications that were simply ineffective. But in many cases, to take medicine in nineteenth-century America was downright dangerous.

The person most responsible for changing this situation—and establishing the beginnings of the FDA as we know it today—was a man named Harvey Washington Wiley. Born in 1844 in a log cabin on the Indiana frontier, Wiley was the sixth of seven children, and grew into a hulking figure, six-foot-one and more than two hundred pounds, with piercing dark eyes and black hair. ("Some said homely," wrote a colleague, "but always most interesting in personality and appearance.") His parents were devoted Christians who, despite their own lack of formal schooling, were committed to providing their children with as much education as possible.

In 1863, the giant farm boy walked five miles on a dirt road to enroll himself at Hanover College. Next, after a stint in the Union army, he got a medical degree from Indiana Medical College and a bachelor's in science from Harvard (which he achieved in only several months), took a side gig as a professor of Greek and Latin, and travelled to Germany, where he studied under some of the top chemists of the time. In 1874, Wiley was appointed as a chemistry professor at the new Purdue University, where he lived up to his playful reputation by buying a nickel-plated Harvard roadster bicycle with a high front wheel and small back wheel and riding it to campus while wearing a "fashionable costume that included knee britches."

Unfortunately, the "infamous bicycle incident of 1880" did not go over well with the buttoned-up Board of Trustees ("Imagine my feelings and those of other members of the board on seeing one of our members dressed up like a monkey and astride a cartwheel riding along our streets," wrote one enraged board member). Wiley decided to leave the university soon thereafter. His next job was with an organization not usually known for its sense of humour: in 1883, he became head of the USDA's Division of Chemistry, the aforementioned precursor to what we now know as the FDA. After spending several years absorbed in sugar research (it is not without reason that he is known as the father of the beet sugar industry), he switched his focus to the adulteration of food.

In 1887, Wiley's department came out with *Bulletin 13*, the first volume of *Foods and Food Adulterants*. Starting with dairy products, it ended up being a ten-part, 1,400-page guidebook, published over sixteen years, that described updated methods for identifying adulterated foods, drugs, and agricultural products. You might not think the public would be eager for scientific reports filled with analyses of baking powders and admonitions against the sale of coppered pickles. But according to one observer, these guides were "in such great demand that they often became unobtainable"—another indication that Americans' modern anxiety about what's in our food may not actually be so modern after all.

Over the course of his research, Wiley became suspicious of many untested chemical preservatives and enhancers being added to foods, from copper sulphate that made old vegetables look greener to borax that improved the smell of questionable canned ham. Many of his suspicions were warranted: the preservatives—which had brand names like Preservaline and the threatening-sounding Freezem—often included ingredients now known to be toxic, several of which have since been banned. But at the time there were no safety-testing requirements for these products, and many of the food companies that bought them claimed to not even know what chemicals they contained, which was quite possible, considering that they weren't required to be listed.

In 1902, Wiley, who by this point had developed a reputation as a strong supporter of the "pure food" cause, received several thousand dollars in congressional funding to investigate his suspicions about preservatives. He used the money to take out an ad in a newspaper to recruit a dozen "young, robust fellows" who were most likely to have "maximum resistance to deleterious effects of adulterated foods." After having respondents agree to participate in the experiments for six months and to absolve the government of any responsibility for what might befall them, he transformed a basement mailroom into a dining hall, and ordered them to consume only the foods and beverages served on its white-tableclothed "hygienic table." He put them through frequent medical exams and made them tote around bags and equipment so that they could collect all of their urine and stools. Then he fed them chemical preservatives and waited to see what would happen.

Wiley started with borax-infused butter, but switched to delivering the chemicals via gelatine caplets after the men complained about the taste. After borax came salicylic acid, sulphurous acids and sulphites, benzoic acid, sodium benzoate, and formaldehyde. Surprisingly, despite the fact that the men knew that they were eating untested preservatives, Wiley had no problem attracting volunteers.

Less surprisingly, he also had no problem attracting press. Newspapers quickly dubbed the group the Poison Squad (slogan: "Only the Brave Can Eat the Fare"), and it wasn't long before the squad was a national sensation, so widely known that even minstrel shows were performing songs about it. To quote the lyrics of one number from Lew Dockstader's minstrel show in October 1903:

> *If you ever visit the Smithsonian Institute,*
> *Look out that Professor Wiley doesn't make you a recruit.*
> *He's got a lot of fellows there that tell him how they feel,—*
> *They take a batch of poison every time they take a meal.*
> *For breakfast they get cyanide of liver, coffin-shaped,*
> *For dinner they get undertaker's pie all trimmed with crepe;*
> *For supper—arsenic fritters, fried an appetizing shade,*
> *And late at night they get a prussic acid lemonade!*

None of Wiley's subjects died as a direct result of the experiments—and there is no record of long-term deleterious effects—but the Poison Squad revealed considerable reason for short-term concern. Too much borax, for example, caused indigestion, severe headaches, and abdominal pain, and eventually left the men incapacitated, producing, as Wiley put it, "an inability to perform work of any kind." (Boric acid is no longer allowed as a preservative.)

Three of the other four chemicals tested in the first round resulted in symptoms that were similar or worse.

By the five-year project's end, Wiley had concluded (as had the members of the Poison Squad) that several of these preservatives were dangerous, and turned his attention toward creating a law that would keep chemical preservatives out of foods. The nascent food industry pushed back, making claims quite similar to those surrounding the issue of supplement regulation today: that the substances weren't as harmful as Wiley asserted, that the government shouldn't interfere with business, and that the would-be reformers were stirring up unnecessary panic.

But several unrelated events soon helped Wiley win the support of the public. In 1905, a muckraking journalist named Samuel Hopkins Adams published the first of an influential ten-part series of articles in *Collier's* magazine that revealed some of the dangers of unregulated patent medicines. It included, among numerous other examples, stories of infant "soothing syrups" whose high morphine content was creating young addicts. Then in 1906, the day after Adams's final article came out, Upton Sinclair's book *The Jungle* was published, in which Sinclair revealed disgusting practices he had observed while working undercover in a Chicago meatpacking plant. Sinclair had hoped for his book to convert readers to socialism, but as he later lamented, he had "aimed at the public's heart and, by accident, I hit it in its stomach." After its publication, American meat sales plummeted.

Without public outrage, Congress would likely have continued to ignore the issue of food and drugs. But this time all the pieces had fallen into place: together the Poison Squad, Adams's articles, and *The Jungle* roused the American public and forced Congress to act. On June 30, 1906, two bills became law—the *Jungle*-inspired Federal Meat Inspection Act (which created the first federal meat inspectors) and the Pure Food and Drug Act. Wiley was personally disappointed when President Theodore Roosevelt gave someone else the pen he'd used to sign the Pure Food and Drug Act, which is often called the Wiley Act in honour of his role in its creation. But that didn't take away from the magnitude of the accomplishment. After more than twenty-five years and a hundred proposed bills (and several years before the discovery of the first vitamins), the United States finally had a federal law on food and drugs.[2]

---

[2] These 1906 bills are the reason that the United States, unlike most countries, has two

While attending a conference about supplement adulteration at the US Pharmacopeia's (USP) headquarters in Rockville, Maryland, I noticed a quotation from Wiley's 1930 autobiography hanging on the wall: "How does a general feel who wins a great battle and brings a final end to hostilities?" Wiley had written. "I presume I felt that way on the last day of June, 1906."

And it's true: the passage of the Pure Food and Drug Act was a watershed moment in food and drug regulation. (It didn't address vitamins or supplements as such because vitamins were just beginning to be suspected, and the modern-day supplement industry did not yet exist.) It required, among other things, that medicines and foods not be adulterated or impure, and that if a company decided to list its product's ingredients on its box, the information must be true. Patent medicines were to be officially counted as drugs, as were any other products that were "intended to be used for the cure, mitigation or prevention of disease" (a phrase still used to define drugs today). The act also banned labels that had "any statement, design or device . . . which shall be false or misleading in any particular."

But as the above description may already suggest, the 1906 Pure Food and Drug Act had many loopholes and ambiguities that were easy to exploit. It didn't require companies to do any sort of quality testing. It specifically listed nearly a dozen "dangerous ingredients" often used in patent medicines, such as heroin, opium, cocaine, and morphine, but allowed them to be included in products as long as they were listed on the box—a requirement that had the unintended consequence of helping addicts to comparison shop. Nonnarcotic ingredients did not have to be listed. There was no clear definition of "false or misleading" (indeed, courts soon judged that quack products' claims that they cured cancer or diabetes were not misleading). The legislation didn't require manufacturers to prove that their products were effective or safe, and penalties for breaking the law were low—it counted as a misdemeanour with a maximum first-offence penalty of $200. And on top of all that, there was a

---

separate agencies regulating its foods and drugs: the Federal Meat Inspection Act gave meat (and later poultry) inspection duties to the US Department of Agriculture, while the Pure Food and Drug Act gave regulatory authority of drugs and non-meat foods to what is now the Food and Drug Administration. This separation has led to some truly odd jurisdictional divides: today the FDA regulates cheese pizza, but the USDA regulates pepperoni pizza; USDA gets spaghetti with meatballs, while the FDA gets meat-flavoured spaghetti sauce with less than 3 percent actual meat.

financial issue: enforcing the law would be expensive, but Congress had not authorized any money to do so.

Unfortunately, Wiley—who retired from the government in frustration in 1912 and went on to work at *Good Housekeeping* magazine—did not live long enough to see the 1906 law updated in any substantial way. He died in 1930, shortly before Walter Campbell, the chief of the agency that later officially became the FDA, made a powerful case in the *New York Times* that the 1906 act needed to be revised. "The weak points of the present law have become increasingly apparent," he wrote, pointing out that only three amendments had been made to the act since its passage more than a quarter of a century earlier, and that the government was supposed to be responsible for "showing affirmatively in every instance that a food containing an added poisonous ingredient may be harmful." He wasn't exaggerating: the number of products on the market that the FDA was supposed to regulate had roughly doubled since 1906, there was a substantial new market for untested chemical-based cosmetics, and the patent medicine market was worth some $350 million, even more than it had been when the Pure Food and Drug Act was passed. In contrast, the FDA's entire annual budget was less than the amount needed by the USDA to produce its newsletter.

To raise public awareness about the issue, the FDA managed to put together a travelling exhibit that publicly highlighted some of the more egregious examples of dangerous products that were allowed under the law, like Lash-Lure, an eyelash dye that blinded many women; Radithor, a toxic radium-laced tonic; and the Diana Ideal Womb Supporter, which could puncture Diana's uterus if it wasn't properly inserted. The exhibit, full of heartbreaking stories, was shown to everyone from Congress members to visitors at the 1933 Chicago World's Fair to unwitting international delegates of the Associated Country Women of the World. The press dubbed it "the American Chamber of Horrors." But it wasn't enough to convince Congress. That took a nationwide tragedy.

Sulfanilamide was one of the world's first anti-infective drugs—it was considered a "wonder drug"—and was used to treat strep throat and other infections like gonorrhea and meningitis. But it was also notoriously difficult to dissolve, and so was only available as a large, bad-tasting pill. Doctors and patients were eager for a liquid form of the drug that would be more palatable, especially to children, and eventually the chief pharmacist and chemist of Tennessee's S. E. Massengill Company came up with a solution: a thick, sweet solvent called diethylene glycol. The company added raspberry flavouring, tested it for taste and smell, and on September 4, 1937, began shipping Elixir Sulfanilamide to stores.

The company later claimed not to have been aware that diethylene glycol, which is related to a main ingredient in antifreeze, can cause kidney failure, convulsions, and a painful and prolonged death—even though there is evidence that Massengill had done preliminary safety testing with diethylene glycol ten months earlier and found that a solution containing 3 percent diethylene glycol caused kidney damage in rats.[3] (Elixir Sulfanilamide contained 72 percent.) Either way, it didn't take long for the tragedies to begin.

By early October, a doctor in Tulsa, Oklahoma, reported that ten of his patients had immediately died after taking the elixir. More quickly followed. On November 1, 1937, a woman from Tulsa named Marie Nidiffer wrote a heartbreaking, handwritten letter to President Roosevelt about the night her six-year-old daughter, Joan, had died. "The first time I ever had occasion to call in a doctor for her and she was given the Elixir of Sulfanilamide," wrote Nidiffer, who enclosed a smiling photo of her daughter with the letter. "Today our little home is bleak and full of despair. . . . Even the memory of her is mixed with sorrow for we can see her little body turning to and fro and hear that little voice screaming with pain and it seems as tho it drives me insane." Within months of the product's launch, an estimated 107 people, mostly children, had died.

The public looked to the FDA to respond, but as the agency's chief, Walter Campbell, was forced to admit at a press conference, the limitations of the 1906 law meant that the FDA couldn't legally do anything to block the product unless there was something wrong with its label. Since the FDA was the only agency with any possible jurisdiction, Campbell decided to start a national investigation anyway, and sent out the FDA's entire field force, then 703 people, to identify and recover every bottle of the elixir—no small task, given that 240 gallons' worth of the product had already been distributed in four-ounce bottles. As the inspectors scrambled, Massengill agreed to issue a recall notice

---

[3] Diethylene glycol has shown up in food products since then, including tainted toothpaste from China in 2007, and wine from several Austrian wineries in 1985.

but didn't mention why the drug was being recalled, or that it was an emergency. The FDA forced the company to issue a more strongly worded warning several days later, by which point many more doses had likely been prescribed.

By the time the Elixir Sulfanilamide deaths were over, the chemist who had come up with the formula had committed suicide. The company would eventually be fined $26,000 (about $240 per death), which was the largest fine the FDA had levied up to that time. But in a letter to the American Medical Association, Massengill's owner tempered his apology. "My chemists and I deeply regret the fatal results, but there was no error in the manufacture of the product," he wrote. "[N]ot once could [the company] have foreseen the unlooked-for results. I do not feel that there was any responsibility on our part."

And though it might sound crazy, he was correct: Massengill was not responsible for proving that the elixir was safe. No one was. In fact, the only way the FDA managed to successfully prosecute the company was through a technicality—according to the 1906 Pure Food and Drug Act, any product labelled as an "elixir" was supposed to contain alcohol, and Elixir Sulfanilamide did not. If it had been called a "tonic" rather than an elixir, wrote the secretary of agriculture, "no charge of violating the law could have been brought."

The flood of publicity and letters caused by the Elixir Sulfanilamide tragedy put pressure on Congress to act, and in 1938 it finally passed the Federal Food, Drug, and Cosmetic Act. It was the strongest such legislation that had ever been enacted in the United States, and fundamentally changed the way drugs were developed and sold. Most important, it shifted the burden of responsibility from the government to manufacturers, who were now responsible for submitting some evidence to the FDA that their products were safe *before* putting them on the market.

Even though the law's proof-of-safety requirements were much less strict than those in force today, and it didn't address efficacy, its impact was substantial and obvious. Whereas in the 1920s the top two hundred drug companies in the United States only had a few scientists on staff, mostly working on chemical processing, by the 1940s there were 58,000 scientists in America's drug industry working specifically on drug research. Part of this increase was due to the advent of penicillin, which helped scientists begin to develop drugs that—unlike most patent medications—actually *worked*. (As journalist

Philip Hilts points out, by the early 1950s, 90 percent of prescriptions were for drugs that hadn't even existed when the law was passed.) Because of these developments—increased regulation high among them—pharmaceutical companies drastically reduced the number of products that they produced. By the 1950s, Smith Kline, for example, had dropped 14,940 of its 15,000 products so that it could focus on the remaining 60.

The 1938 law was a crucial step toward the pharmaceutical testing requirements that we take for granted; even today, more than seventy-five years later, it still remains the backbone of America's regulation of foods, drugs, and cosmetics. However, the law said next to nothing about vitamins, the first synthetic versions of which had been brought to market shortly before the act was passed. Nor could it say anything about the more exotic types of supplements that now cling to vitamins' coattails, because they didn't yet exist.

Had vitamins been discovered and synthesized earlier, they might have been addressed more explicitly in the 1938 Food, Drug, and Cosmetic Act. This in turn might have set a precedent that could have been applied to other dietary supplement products, and today's supplement market might look radically different as a result. Instead, as American food and pharmaceutical regulations became increasingly strict, the government oversight of America's vitamins and supplements took a different path. For a while, it simply stayed stuck in time. And then it began to move backward.

When I got home from my dermatological expedition to GNC, I immediately opened my bottle of sea buckthorn/omega-7 capsules and shook several onto my palm. They were dark and—in a welcome change from the fishy stench of the omega-3s I'd been taking—smelled vaguely of licorice. I bit into one for the sake of journalism and it exploded into a bitter oil that coated my tongue, leaving me with a deflated capsule in my mouth and a resolution never to try that again.

Unlike that of a prescription or over-the-counter drug, the box that contained my sea buckthorn oil didn't have a patient information insert revealing its chemical structure, what its side effects might be, or a clear description of what it was supposed to do. Instead, a Google search taught me that sea buckthorn's primary omega-7 fatty acids are cis-vaccenic and palmitoleic acid, the former being particularly prevalent in dairy foods and the latter in macadamia nuts (and, of course, sea buckthorns). Unlike omega-3 and omega-6 fatty acids, omega-7s are not essential—that is, our bodies are able to make them; the "Supercritical" on the label merely refers to the company's extraction process ("Modern science has determined how best to extract Sea Buckthorn's broad spectrum constituents," it nebulously explains). I learned that Omega 7 is the name of a Cuban paramilitary group, found legends claiming that sea buckthorn berries were responsible for the success of Genghis Khan, and read several disturbing suggestions that consuming too many omega-7 fatty acids might make me develop "old person smell."

Missing from the many product descriptions of various brands of sea buckthorn, however, were references to any specific scientific studies that could back up the box's therapeutic claims. The closest I found was a statement on the company's website about a 1999 paper that "suggested that this palmitoleic acid may help support healthy skin"—but in addition to its wishy-washy phrasing, it didn't name the trial or say where it had been published. Nor did the company give any indication of its sourcing, manufacturing, or quality assurance practices.

I hoped the FDA might have further information, but when I visited its website, I found this disclaimer in its Q&A on dietary supplements: "FDA does not keep a list of manufacturers, distributors or the dietary supplement products they sell. If you want more detailed information than the label tells you about a specific product, you may contact the manufacturer of that brand directly."

Surprised, I checked the buckthorn box for contact information. Luckily it did list the company's name, address, and phone number (which is required by the FDA, although not all companies comply). I found my sea buckthorn in the company's online product catalog, which was presented as a fairytale-like book on the backdrop of a mossy forest floor. But the site provided no substantiation for the claims on the box, so I called the customer support line.

"I'm thinking of taking sea buckthorn oil and was wondering what studies have been done on it," I asked the customer service representative who came on the line. "I'm trying to find out more about how it works."

"Well, we're not a research organization, so we don't actually produce any reports on the subject," he told me.

"I know," I reassured him, as an animated bug crawled across my screen. "But there's a specific 1999 study that's mentioned on your website. Do you know where I can find it?"

He did not. "You can see on the label that it's not evaluated, so we like to let people know that the info's out there, but we're not actually able to produce it," he offered by way of explanation. "You might want to look up sea buckthorn on PubMed"—the online database of biomedical journals provided by the National Library of Medicine at the National Institutes of Health (NIH). "I'll take a quick pass and see if I can find anything."

I waited as he did a search. "There are quite a few clinical abstracts—I'm showing eighty-three," he said. "But I'm not seeing that 1999 study. You might want to inquire at your local library to see if they have any more information."

I was not reassured by this turn of events: the government directing me to the supplement company, which directed me back to the government and, when that failed, sent me to everyone's trusted dietary supplement expert, my local librarian. When I asked him why the box said women weren't supposed to take sea buckthorn while pregnant, his answer was equally unsatisfying.

"We can only suggest a handful of our products during that cherished time of life," he said, in a cadence that suggested that for him "cherished time of life" was as common an expression as "beg your pardon" or "have a nice day." He continued, "It's not a safety issue, but there's not a lot of research on herbal products and their use during pregnancy. It's not that we're aware of any issues with sea buckthorn; we just want to play it safe."

If sea buckthorn was sold as a prescription or over-the-counter drug rather than a supplement, the representative's answers would have been quite different. In fact, I probably wouldn't have called him in the first place, because there would have been information on the label or a patient information insert answering my questions and providing evidence that the buckthorn was both effective and safe.

Over-the-counter and pharmaceutical drugs haven't always had to include this type of information on efficacy and safety—its availability is in large part due to the 1962 Kefauver-Harris Amendments, which were passed unanimously by both houses in response to the horrible birth defects caused by thalidomide, a morning sickness drug.[4] The amendments required that drugs be proved safe and effective by "adequate, well controlled" studies (the 1938 law hadn't addressed efficacy and had left safety standards vague) and eventually led to the rigorous drug approval process we have today—which in turn has made America's pharmaceutical market the most trusted in the world.

As for the safety of America's *food* supply, a series of laws and regulations since 1938 has improved it as well, including requirements for pre-market testing of new food additives and colourings, regulations regarding sanitation, good manufacturing practices, labelling and quality control, programs to track down the source of outbreaks of food-borne illnesses, and the creation of a list of ingredients that are "generally recognized as safe" (GRAS).[5] New ingredients that are *not* listed as GRAS are supposed to require pre-market safety testing and approval, though the process is nowhere near as costly or time-consuming as that for drugs. Despite Americans' constant argument about the proper size of government, over the past century we seem to have reached the general consensus that some degree of regulation is necessary to keep our drugs and food safe.

We've also agreed that drugs should require more pre-market testing and FDA approvals than food. This makes sense, given that most food safety concerns are about events that occur during or after production, like contamination and spoilage, not about the raw ingredients themselves. But this distinction between foods and drugs—or, more specifically, the differences in how they're regulated—leaves an enormous middle ground: the products that could be considered foods *or* drugs.

Take vitamin C as an example. It prevents and cures scurvy, a miraculous ability that, if it had originally been a man-made creation, would classify it as

---

[4] The FDA never approved thalidomide for sale in the United States, but 20,000 samples of it were distributed anyway. It produced far more birth defects in Europe, where it had been approved for sale in several countries.

[5] The GRAS list also includes many additives and chemicals that were in use at the time of the list's creation and were grandfathered in without any clinical trials—but that's a whole different controversy.

a drug. (And an invaluable one, considering that we'd die without it.) But vitamin C is abundantly available in oranges and fresh produce—substances that are unquestionably food. So how should it be classified? What rules should apply to vitamins and their nutritional companions, minerals—let alone the ever-expanding array of other supplements on drugstore shelves?

By the time the 1938 Food, Drug, and Cosmetic Act was passed, a number of state legislatures had decided that synthetic vitamins—which, again, were at the time relatively new products—should be classified as drugs, to be sold only in pharmacies and often by prescription. But the 1938 federal act came to a different conclusion: while it only mentioned vitamins once by name, the implication was that as long as they weren't prescribed for illness or sold with disease-preventing health claims, vitamins should be considered foods. In 1941, the FDA issued regulations regarding how vitamin and mineral supplements should be *labelled*, but didn't set any restrictions on what ingredients (or in what quantities) they could contain.

If the vitamin and supplement market had remained capped at 1941 levels, the FDA might never have seen a need to further intervene. Instead, it grew steadily through the 1940s and 1950s and then exploded during the 1960s, as general distrust in government increased and so-called alternative medicine and natural foods began to go mainstream. Vitamins in particular got celebrity boosts from Adelle Davis, an enormously influential (if often scientifically irresponsible) nutritionist who advocated high levels of synthetic vitamins,[6] and Linus Pauling, the double Nobel Prize-winning chemist who claimed that high doses of vitamin C could cure everything from the common cold to cancer—both incorrect hypotheses that many people still believe today. Between 1972 and 1974, American sales of vitamins, minerals, and other supplements increased from approximately $500 million to $700 million, a 40 percent rise.

---

[6] Davis, who had a master's degree in biochemistry from the University of Southern California, projected an image of scientific rigor. But she was prone to hyperbole, such as her claim that "your nutrition can determine how you look, act, and feel; whether you are grouchy or cheerful, homely or beautiful, physiologically and even psychologically young or old; whether you think clearly or are confused, enjoy your work or make it a drudgery, increase your earning power or stay in an economic rut." She also sometimes played fast and loose with her interpretations of others' work: when a researcher examined the 170 citations in one chapter of her 1965 best seller, *Let's Get Well*, he found that only 30 of them actually supported Davis's claims.

Aware of the public's increasing enthusiasm for vitamin supplements, the FDA became concerned. For while it would be extremely difficult to overdose on most vitamins at the levels found naturally in food, it is possible to do so with pills—especially the fat-soluble ones (A, D, E, and K) because they don't dissolve in water and thus aren't easily excreted. Vitamin A is by far the most dangerous, since too much vitamin A can cause irreversible liver damage, birth defects, and in extreme cases, death. In one famously dramatic example, an early twentieth-century Antarctic explorer died of vitamin A poisoning after eating several of his sled dogs' livers. (Huskies, as well as seals, polar bears, and walruses, store extraordinarily high levels of vitamin A in their livers—an important factoid if you're planning a polar expedition.) As for the other fat-soluble vitamins, high levels of vitamin E can interfere with blood clotting, vitamin K can interfere with blood-*thinning* medications like warfarin, and routinely consuming too much vitamin D can lead to dangerously high blood calcium levels and eventually cause calcium deposits in places you don't want them, like your arteries or kidneys.

But while the fat-soluble vitamins pose the most dramatic risks, high doses of water-soluble vitamins can cause side effects as well. Too much niacin can lead to an uncomfortable and itchy condition known as niacin flush, as well as cause interactions with cholesterol-lowering drugs and increase the risk of liver damage. High levels of folic acid (B9) can mask signs of a deficiency in vitamin B12, which, if left untreated, can cause irreversible cognitive impairment. Even vitamin C is not always safe; at very high doses, or when it's not properly excreted, it can cause side effects ranging from severe diarrhoea to kidney stones. (And this says nothing about minerals, which can also be toxic in high doses.)

If you think about it, it makes sense that higher-than-normal amounts of vitamins might do additional, and potentially unwanted, things in our bodies. As *The Mount Sinai School of Medicine Complete Book of Nutrition* explains, "Vitamins and minerals that are in excess of those needed to saturate enzyme systems function as free-floating drugs (instead of as receptor-bound nutrients) . . . and like all drugs, they have a potential for adverse side effects."

Again, husky livers aside, vitamins pose no danger for most people at levels naturally found in food, and taking a multivitamin every day is not going to cause an acute overdose. (And, as noted, there are millions of people in the developing world who suffer from serious micronutrient deficiencies, for whom supplements and fortified foods are truly lifesaving.) But when influential public figures like Linus Pauling and Adelle Davis began encouraging the idea that megadoses of vitamins (and minerals) could be not just safe but *beneficial*, the FDA's anxiety increased. What if, in response to public demand, manufacturers began producing vitamin tablets in potencies so high that they were actually dangerous?

The agency had issued proposed regulations for vitamin and mineral supplements in 1966, which included the suggestion that multivitamin and mineral product labels bear a statement that read "Vitamins and minerals are supplied in abundant amounts in the foods we eat. . . . Except for persons with special medical needs, there is no scientific basis for recommending routine use of dietary supplements." This statement in particular was met with what former FDA commissioner David Kessler referred to as "uniform disapproval" (including from many nutritionists). Thanks to successful lawsuits filed by industry groups, millions of public complaints instigated by a supplement-industry lobbying group called the National Health Federation, and the inherently long process of issuing new regulations, the FDA did not publish its final regulations—which didn't contain the statement—until 1973.

These proposed regulations attempted to standardize vitamin and mineral products and limit the combinations of vitamins and minerals that dietary supplements could contain. They also stated that any single pill that contained more than a certain percentage of that vitamin's RDA (150 percent for most vitamins) would be treated as an over-the-counter drug—meaning it would require the same sort of proof of safety and efficacy as something like ibuprofen. In addition to limiting the chance of accidental overdoses, the regulations would have given the FDA a simple way to determine which products were out of compliance with government regulations, rather than having to go to court for every case.

It's worth noting that the FDA's regulations would not have *forbidden* vitamins from being sold in higher potencies; they simply would have required any product that exceeded the FDA's limits to be backed with evidence of safety and efficacy similar to what we take for granted every time we buy antibiotic ointment or an allergy medication (or, for that matter, antiperspirant or sunscreen lotion, both of which have active ingredients that are considered over-the-counter drugs). Consumers who wanted to take higher doses could have bought the approved higher potency formulation or swallowed additional pills. And companies who wanted to create new formulations could also have done so, as long as they could prove their products were effective and safe. Given the past half century's march toward tighter regulation of the food and drug market, these rules might not seem radical—in fact, you could argue that they made good sense. So why is it, then, that they never actually went into effect?

Before delving too deeply into the reaction to these proposed regulations, let's take a step back for a moment to acknowledge how the FDA works and the limitations of its powers. The FDA has an enormous range of responsibilities. By its own estimates, it regulates roughly 25 percent of every consumer dollar spent in the US economy, including food, pharmaceuticals, cosmetics, veterinary and tobacco products, medical devices, and blood transfusions. But contrary to public belief, the FDA—which is now an agency of the US Department of Health and Human Services—doesn't choose what it regulates; nor does it have exclusive control over *how* it regulates these products. Instead, its job is to implement laws passed by Congress, which means, to put it more bluntly, that Congress tells it what to do. As Daniel Fabricant, former director of the FDA's Division of Dietary Supplement Programs, explained, paraphrasing the Tennyson poem "The Charge of the Light Brigade": "Mine is not to wonder why; mine is but to do or die."

In practice this means that there are only two situations in which the FDA can issue new regulations: (1) if the proposed new rules are in an area of law that the FDA already regulates, in which case it must follow an official rule-making process that can take years, or (2) if Congress passes a new law telling it, or granting it the authority, to regulate something new. Conversely, if Congress passes a law that says that the FDA *cannot* regulate something, the agency must also comply. In any case, the fact that the FDA's rule making is subject to multiple rounds of public comments means that even when it's trying to issue rules in an area over which it definitely has authority (such as food, drugs, and supplements), its proposed rules can be significantly influenced by interested parties—including the supplement industry.

In this case, the supplement industry responded to the FDA's regulations with another lawsuit challenging the FDA's authority. And again, its effort was successful: the court ruled that the FDA did not have the authority to standardize vitamin and mineral supplements, or to classify high-potency supplements as drugs just because they were potentially toxic or nutritionally unnecessary.

However, the supplement industry was not satisfied. Instead, the afore-mentioned National Health Federation began lobbying for something much more dramatic: a bill that would *permanently* limit the FDA's ability to set limits on the potency of vitamin pills. The idea seemed so ludicrous that, as journalist Dan Hurley describes in his book *Natural Causes*, a House staffer knowledgeable in health legislation summed up its chances for passage with two words: "No way."

But about a year later, Senator William Proxmire (Democrat from Wiscon-sin), a politician best known for criticizing wasteful government spending via his tongue-in-cheek Golden Fleece Awards, introduced a bill that would do exactly what the NHF had requested—and then some. The Vitamin-Mineral Amendment, which is now commonly known as the Proxmire amendment, would forbid the FDA from *ever* limiting the potency of vitamin and mineral pills or classifying them as drugs. It would prevent the FDA from requiring that supplements contain ingredients proven to be useful, or banning the inclusion of ingredients known to be use*less*. It would also forbid the FDA from ever establishing standardization requirements for supplements of any kind.

Far from being concerned about this dramatic reduction of oversight, the public reacted so positively to Proxmire's proposed bill that a bipartisan group of forty-four other senators signed on as cosponsors. It was one of the first concrete examples of how powerful Americans' emotional attachment to vita-mins had become, and how this attachment could be harnessed as a potent political force—one on which the supplement industry has continued to capi-talize ever since.

The congressional hearings for the Proxmire bill began on August 14, 1974. They were presided over by Senator Edward Kennedy (Democrat from Massachusetts), who began his argument against the bill by stating that the FDA had the responsibility to protect American consumers from foods and drugs that were "potentially harmful to their health," as well as from products that were advertised to be "therapeutic or in some other way beneficial when, in fact, they may be worthless and a waste of money."

The FDA, whose position was backed by numerous groups, including the AARP, the American Academy of Pediatrics, and the American Society of Clinical Nutrition, put up a powerful case of its own. Its then commissioner, Alexander Schmidt, MD, entered into the record a seventy-two-page single-spaced review of actions the FDA had taken against supplement makers between 1960 and 1962, and pointed out—in a statement still relevant today—that "what is overlooked by a great many people is that while there is not a lot of evidence that very large doses of water-soluble vitamins are harmful, there is not a lot of information that large doses of water-soluble vitamins are safe, either."

The FDA also called upon Sidney Wolfe, MD, a doctor who had created the health arm of Ralph Nader's consumer advocacy group, Public Citizen. "This is a drug industry," Wolfe said. "The differences between large doses of vitamins . . . and over-the-counter [drugs] are nonexistent as far as I'm concerned."

A lawyer with Nader's organization, Anita Johnson, came to the same conclusion. "Everything we have learned from environmental hazards shows us you cannot assume a product is safe if it has not been tested," she said, claiming that the amendment would "decimate" the FDA's ability to regulate supplements. Johnson also pointed out an extremely important, and prescient, aspect of the amendment: it would apply not just to vitamins and minerals, she said, but to "special dietary foods and all ingredients of special dietary foods."

The hearing's most visually memorable presentation was undoubtedly that of Marsha N. Cohen, an attorney with Consumers Union, the policy and action division of *Consumer Reports*, the nonprofit that independently tests and rates consumer products.

"Why should vitamin and mineral supplements be regulated? I have here with me a little visual demonstration of our answer to that argument," she said as she brought out her evidence: eight cantaloupes, which she arranged on the table before her. "We can safely rely upon the limited capacity of the human stomach to protect persons from overindulgence in any particular vitamin or mineral-rich food. For example, you would need to eat eight cantaloupes—a good source of vitamin C—to take in barely one thousand milligrams of vitamin C. But just these two little pills, easy to swallow, contain the same amount. . . . If the proponents of the legislation before you succeed, one tablet could contain as much vitamin C as all these cantaloupes—or even twice, thrice, or twenty times that amount. And there would be no protective satiety level."

But as dramatic as these presentations were, Proxmire was ultimately more effective. In his argument, he made the valid point that the RDAs had already been changed and updated numerous times since their creation, and that there were many nutritional experts who thought that the recommendations were still too low. How was the industry supposed to be compliant if the FDA constantly updated the definition of what potency was permissible? He also argued that if vitamins were to be regulated like drugs because of their potential toxicity, it could open the door for the FDA to set limits on *any* food or food ingredient that is toxic in high doses, a category that includes salt, caffeine, and even water.

Perhaps the most ingenious aspect of Proxmire's argument was the way he framed the issue of supplement regulation. "What the FDA wants to do is to strike the views of its stable of orthodox nutritionists into 'tablets,' and bring them down from Mount Sinai where they will be used to regulate the rights of millions of Americans who believe they are getting a lousy diet to take vitamins and minerals," he said. "The real issue is whether the FDA is going to play God."

The argument wasn't about safety, in other words. It wasn't about efficacy. It was about whether rule-obsessed, megalomaniacal bureaucrats should be allowed to limit Americans' right to make decisions about their own health. This argument—that access to vitamins is a matter of personal freedom—was a brilliant tactical move, and it has defined the discussion on supplement regulation ever since.

The Proxmire amendment was tacked onto a health bill and passed the Senate with a vote of 81–10. Enacted on April 23, 1976—and still in effect today—the amendment made it illegal for the FDA to *ever* establish standards for supplements, classify them as drugs, or require that they only contain useful ingredients. It forbade the FDA from ever setting limits on the quantity or combination of vitamins, minerals, or other ingredients that a supplement could contain, unless the FDA could prove (usually after the product was on the market) that the formulation was unsafe—an extremely important shift of responsibility. In so doing, the Proxmire amendment brought the commercialization of products like my sea buckthorn oil—as well as the enormous selection of other dietary supplements available today—one step closer.

The commissioner of the FDA, Alexander Schmidt, called the Proxmire amendment "a charlatan's dream." The supplement industry, on the other hand, credited its passage for "the survival of our industry. . . . [W]ithout that, the FDA . . . could have crippled us." Regardless of whose interpretation you agree with—or, for that matter, if both are correct—one thing is clear: the passage of the Proxmire amendment marked the first time that Congress had amended the 1938 Food, Drug, and Cosmetic Act to *withdraw* powers from the FDA. And thanks to the rising power of the supplement industry, it wouldn't be the last.

8

The People's Pills

Recognize at the outset that the dietary supplement
industry is essentially unregulated. When consumers pick up a
dietary supplement today, they assume that the product is safe.
But the fact is, there has never been a systematic evaluation
of the safety of dietary supplements. And, when consumers
see a health claim for a dietary supplement, they assume
it will provide the benefit it touts. In fact, the marketplace
is awash in unsubstantiated claims. We have a
serious problem on our hands.
—DAVID KESSLER, MD, COMMISSIONER OF
THE FOOD AND DRUG ADMINISTRATION, AT
CONGRESSIONAL HEARINGS FOR THE DSHEA, 1993

As the Proxmire amendment makes clear, vitamins are hardly blameless bystanders in the history of America's supplement regulations. Instead, vitamins opened the door for a whole raft of questionable products; my Supercritical Sea Buckthorn oil—like most of the products in GNC—owes its very existence to vitamins' impact on regulation and the public's embrace of them as miracles in pill form. After all, it was vitamins that first created a conceptual bridge between foods and pharmaceuticals. It was vitamins that introduced and popularized the idea that substances found naturally in foods could have astounding effects on health and that you might need to supplement your diet with pills. It was vitamins' relatively low risks that inspired the public to assume that *all* supplements must be safe. And it was vitamins that mobilized the public to rally in support of the Proxmire amendment to begin with. Vitamins may appear to be as innocent as oranges, but when it comes to their effects on regulation, they're the equivalent of gateway drugs.

In the late 1980s, some fifteen years after the passage of the Proxmire amendment, a strange and frightening illness emerged that tragically demonstrated the potential dangers of limiting the FDA's authority to regulate

supplements. Now known as eosinophilia-myalgia syndrome (EMS), the illness occurred among people taking dietary supplements of an amino acid
called L-tryptophan that had gained popularity as a treatment for a number of
problems, including premenstrual syndrome, depression, children's attention
deficit disorders, and most commonly, insomnia. Like vitamins, L-tryptophan can be found naturally in food, and so the supplements had been advertised as natural and safe—even though they had been on the FDA's radar as a
possible problem for several years. Thanks in part to the Proxmire amendment, the FDA had not been allowed to require any sort of pre-market testing
of L-tryptophan for safety or efficacy; nor were there required good manufacturing practices or post-market batch testing.

Dorothy Wilson, a healthy and active woman who worked as a manager at
Unisys in Philadelphia, was one of the many Americans who had begun taking L-tryptophan to help her sleep—and like most other victims, she did not
make an immediate connection between her terrifying symptoms and the
L-tryptophan that had been recommended to her by her doctor. As she
described her experience at a congressional hearing in 1993:

> After four months of L-tryptophan, I felt a strange sensation in my legs.
> Painful muscle spasms attacked my body. Tests showed elevated enzymes,
> high white blood count, and other abnormalities, plus eosinophilia. My
> doctor knew of no illness with these symptoms.
>
> Soon I had trouble getting up from chairs, walking stairs, my menstrual
> cycle stopped, body hair thinned. I lost my appetite, I fell frequently.
> Weakened dramatically, I was hospitalized. An EMG and nerve biopsy
> showed extensive nerve damage. A cancerous breast mass was found. . . .
> [A]n itchy rash appeared. I had fever, night sweats, my skin became hard
> and tight. Bedridden, I had to be moved from a lying position to a sitting
> position. Bed sores developed. I gave power of attorney to a friend, I could
> not sign my name. My jaw locked. I had difficulty eating. My voice weak-
> ened. I couldn't cough or sneeze. . . .
>
> After inpatient rehabilitation and years of outpatient therapy, I spend
> my days in a wheelchair. . . . I use steroids, spasm-controlling Serax, Per-
> coset and morphine. A year after I stopped L-tryptophan, bladder weak-
> ness developed. This year thyroid and pituitary gland problems. New

symptoms and relapses never stop. Still, I am fortunate. I do not yet have
brain lesions with confusion and memory loss.

Wilson told the audience that, despite the physical challenges of getting to
the hearing, she had a "burning desire" to be present. "I wanted you to look
at me condemned to a life of severe pain and disabilities," she said, "and be
reminded of the price I pay every day of my life due to L-tryptophan."

Today, there is no doubt that Wilson's supplements caused her medical
problems, but at the time the relationship was not immediately obvious.
Instead, one of the first people to suspect the connection was a young reporter
at the Santa Fe bureau of the *Albuquerque Journal*, who'd been given an assignment to look into the cases of two New Mexico women who had arrived at the
hospital with similar mysterious symptoms.

The symptoms were strange enough that the reporter, Tamar Stieber,
soon began to uncover other cases, and she established a link between the
victims: they'd all been taking L-tryptophan supplements. Several researchers at the National Institutes of Health made a similar observation, and the
FDA eventually called for a voluntary recall of all L-tryptophan in the country. In the end, nearly all cases were linked to L-tryptophan made by a particular Japanese producer, which had made changes to its purification methods
shortly before the outbreak occurred that are thought to have resulted in the
inclusion of an EMS-causing contaminant.

The company, Showa Denko, eventually paid about $2 billion to victims.
But as of 1998, when another form of tryptophan began causing problems, no
one had conclusively identified what the original contaminant had been, or
why particular people had more serious side effects than others. The chief of
environmental hazards at the Centers for Disease Control and Prevention
(CDCP) announced that *sixty-three* different contaminants had been found in
the suspect batches of L-tryptophan, and the exact cause of EMS, which
killed an estimated forty people and affected more than fifteen hundred, was
never determined.

The fact that a contaminant was likely to blame in most cases of EMS
does not mean, however, that pure L-tryptophan itself would necessarily
be problem-free. "[I]t's not just impurities that cause these compounds to
be unsafe for some people," said Richard Wurtman, MD, a professor of

neuroscience at Harvard Medical School and the Massachusetts Institute of Technology, at the 1991 congressional hearing on the tragedy. Wurtman, whose lab had published some four hundred papers about tryptophan and amino acids, explained to the audience that pure L-tryptophan can interact dangerously with a number of psychiatric and cardiovascular drugs. "Tryptophan was," he said, "in every sense, an accident waiting to happen."

The reason, Wurtman explained, is that to our bodies, an isolated amino acid is not a food—it's a drug. The fact that tryptophan naturally occurs in foods like turkey does not mean that it is necessarily safe in supplement form.

"Tryptophan in dietary protein is an important nutrient," said Wurtman. "When you have it in protein it comes along with twenty-one other amino acids, and you need the pattern, all of them, in order to utilize them to make your own protein. When you take pure tryptophan in pills or a bottle, *it's not natural.* Never in man's evolutionary history did he or she take an individual amino acid of that sort. . . . The body does not handle it the same way it handles tryptophan in protein. . . . So tryptophan, in spite of being called a nutritional supplement, has nothing whatsoever to do with nutrition. Tryptophan is a drug . . . its administration in pills does change the chemistry of the brain. [Emphasis mine.]"

At the congressional hearing for L-tryptophan, Dorothy Wilson read aloud the label of a bottle of L-tryptophan that didn't mention any of the potential problems from which she now suffered. "Why didn't the FDA require a warning of possible side effects?" she asked. "I am irreversibly paralyzed. The excruciating pain, spasms, electric shocks, burning and aching muscles have grown worse. I cannot work and am restricted to a house with thirteen steps. I suffer from exhaustion, weakness and muscle fatigue, which often makes it impossible to undress, transfer from the wheelchair, or even move the wheelchair by myself. I need an aide to help me shower and do the basic things that most people take for granted. And the story's obviously not finished."

The L-tryptophan disaster might not have occurred if the FDA had had the authority to require pre-market clinical trials of L-tryptophan, or to review the product's safety record and manufacturing practices *before* it went on the market. Not only would that potentially have spared Dorothy Wilson, but it would also have had another often overlooked effect highlighted by Wurtman: it would have preserved the possibility that L-tryptophan could be developed into a useful drug. After all, many of the medications we take today were derived from chemicals found in nature, including aspirin (the active ingredient of which occurs naturally in willow bark), digoxin (a type of heart medication that is derived from foxgloves), morphine, and antibiotics. At the hearing, Wurtman expressed remorse over the loss of what he believed could have been a safe and effective treatment.

"[N]obody argues about whether or not tryptophan works. . . . It's an effective compound [that] does not cause amnesia or related side effects," he said. "[M]y associates and I proposed . . . that perhaps someday tryptophan would become a legitimate drug that could be used to help people sleep, diminish pain, control mood and appetite, and so forth.

"I had assumed . . . that pharmaceutical companies might take this discovery and invest the ten or twenty million dollars, whatever it took them, to do appropriate safety and efficacy studies," he continued. But, as was by that point all too obvious, "[i]t didn't work out that way."

Instead, L-tryptophan had been bottled and sold as a supplement, with none of the safety, dosing, and efficacy research—or good manufacturing practices—that would have been required of a drug. Tragedy had ensued, L-tryptophan had become a dirty word, and as a result, the chance to develop it into a safe and useful medication—one that might well have helped many thousands of people—had been lost.[1]

In the past, the type of avoidable tragedy caused by L-tryptophan had served as a trigger for Congress to pass legislation requiring the FDA to tighten its regulations, as had happened with the Elixir Sulfanilamide disaster in the 1930s, and thalidomide, the morning sickness medication that caused devastating birth defects, in the 1960s. In the case of L-tryptophan, the target became the health claims that companies were using to sell their foods and supplements. The logic was that the claims being made for L-tryptophan had

---

[1] Imagine, for example, that penicillin mold—one of the world's first and most important antibiotics—had been sold as a dietary supplement and that contaminated batches had begun to kill people. If that happened today, it's possible that the resulting stigma would prevent penicillin from ever being pursued as a legitimate drug. Also, why would a drug company invest time and money into developing a drug if people could already buy its active ingredient as an over-the-counter supplement?

encouraged wider use of the supplement, and therefore the claims had
increased the number of people who'd been harmed.

At the time of the L-tryptophan/EMS incident, health claims were one of
the few areas of the supplement market over which the FDA still had considerable control. Following the passage of the Proxmire amendment, in 1979 the
FDA had withdrawn the entire set of vitamin and mineral supplement regulations that it had been working on for the previous two decades, essentially
giving up on the idea of regulating the supplement market as a whole instead
of on a case-by-case basis. This withdrawal meant that while supplement
makers were free—as they still are today—to sell vitamins and minerals and
other supplements in whatever dose and combination they wanted (which is
part of the reason that L-tryptophan was able to be sold to begin with),[2] they
were not allowed to make health claims for their products. If a company
claimed that its supplements (or foods) could prevent, treat, mitigate, or cure
any disease, those products were to be classified—and regulated—as drugs.
There was supposed to be no middle ground.

But as the marketing for L-tryptophan made clear, there was actually an
enormous middle ground—and its exploitation had been instigated by a seem-
ingly innocent player: Kellogg's breakfast cereal. In 1984, based on a decision
by the National Institutes of Health, the company had begun selling boxes of
All-Bran cereal with labels suggesting that high-fibre cereals (including, of
course, Kellogg's All-Bran) might reduce the risk of cancer. The FDA was not
pleased, but the tactic proved effective—a later analysis by the FDA found
that All-Bran's market share had increased by 47 percent within the first six
months of the campaign.

Kellogg's competitors had pressed the FDA for a response, and the agency
had responded by saying that it would give food companies a "cautious green
light" to make health claims as it worked on developing official rules. Several
years of negotiations ensued, during which time food companies participated
in a health-claim free-for-all. According to one industry estimate, more than
40 percent of the new food products introduced in the first half of 1989 were
labelled with general and specific health claims.

These products' profits did not go unnoticed by the rest of the food and

---

[2] Technically, the FDA can pull vitamin products from the market if they pose a risk to
public safety—but it only has this power after the product has come to market.

supplement industry. The Council for Responsible Nutrition (CRN), one of the supplement industry's leading trade associations, had accused regulatory agencies of having a "not-for-supplement bias," and supplement makers had begun to make health claims for their products as well—L-tryptophan included—while the FDA continued to work on its guidance.

But the FDA's rule-making process is slow, and the L-tryptophan incident brought new urgency and public pressure to the issue of health claims on food products. So Congress stepped in, passing the Nutrition Labeling and Education Act of 1990 (NLEA) before the FDA had finished its own rule making on the subject. The act mostly had to do with creating standardized nutrition labels, and led to the Nutrition Facts panels that now grace most packaged foods. And in an attempt to prevent future incidents like L-tryptophan, Congress also used the act to step into the debate over health claims. It required that health claims on foods be authorized by the FDA and that every claim be supported by "significant scientific agreement" among "qualified experts"—a high, if largely undefined, standard still in effect for food today.

The law also required *supplement* health claims to be regulated by the FDA. But the details of this regulation proved trickier to resolve, in large part because of the influence of Senator Orrin Hatch (Republican from Utah), whose state is home to the so-called Silicon Valley of the supplement industry. At the time, supplements were Utah's third-largest industry, bringing in more than $3 billion a year in sales; as of 2012, they were the state's *largest* industry, at $7.2 billion. "Without Senator Hatch we would not have been able to grow the business," said the executive director of the United Natural Products Alliance in 2012. As Marc Ullman, a lawyer for several supplement companies, phrased it to the *New York Times* in a 2011 article, "he's our natural ally."

Hatch himself had sold vitamins as a young man and has been reported to take a pack of supplements every morning; he has also received hundreds of thousands of dollars in political donations from supplement companies, and owned a small stake in a dietary supplement business. Thanks in part to Hatch's influence, Congress punted on the issue of supplement health claims: the 1990 Nutrition Labeling and Education Act required that supplement health claims be regulated, but left the task of actually *writing* those rules to the FDA.

To the industry, this may have seemed like a strategic move, given that the FDA at the time was in disarray. But it didn't take into account the personality of its new commissioner, David Kessler, who happened to be sworn in on the same day that President George H. W. Bush signed into law the 1990 Nutrition Labeling and Education Act. Kessler, who already had medical and law degrees, was appointed with bipartisan support—including that of Orrin Hatch, with whom Kessler had worked before. But Kessler, vowing that the FDA would no longer be what he called a "paper tiger" about the enforcement of America's food and drug laws, soon made it obvious that he wasn't going to be an industry pawn.

Under Kessler's leadership, the FDA soon issued proposed rules stating that supplement health claims should have the same requirements as food—meaning that they should also be backed by "significant scientific agreement." The logic of this decision was simple: Why should a vitamin D claim on a box of cereal be held to a different standard as the same claim on a supplement bottle—especially given that the supplement industry itself had long argued that its products should be treated as foods? But as Kessler described it to me, "It set off a firestorm."

The firestorm was started by one man in particular, whose last name also happened to be Kessler. Gerald Kessler, to be exact, the founder and sole owner of Nature's Plus, which by the 1990s was one of the top ten supplement manufacturers in the United States—and which made Gerald Kessler one of the richest men in the industry. Gerald Kessler recognized the potentially disastrous consequences that the proposed health-claim rules would have on the supplement industry, since they would require supplement makers to provide some proof that the products they were selling actually did what they claimed to do. He also feared the passage of a pending bill, introduced by Representative Henry Waxman (Democrat from California, and a longtime critic of the supplement industry), that would give the FDA subpoena power to investigate potential violations of the Food, Drug, and Cosmetic Act, as well as enable both the FDA and district courts to order recalls if the violation "involves fraud or presents a significant risk to human or animal health"—a detail clearly intended to help the FDA go after supplement claims.

So in February 1992, within months of the publication of the FDA's proposed health-claim rules for supplements, Gerald Kessler persuaded several dozen industry leaders to convene at the 17,000-square-foot lodge of his 240-acre Circle K Ranch (which had originally been built for Ray Kroc, the founder of McDonald's) for what journalist Dan Hurley calls "a war council unlike any before or since in the history of the supplement industry."

In addition to representatives from the top supplement companies in the country, two of Senator Hatch's top aides were present. While Gerald Kessler's colleagues were initially skeptical, he eventually convinced them of the impending threats to their businesses and the need for the industry to take aggressive action. The group decided to form an umbrella organization of supplement industry groups, the Nutritional Health Alliance (NHA), which would enable them to use their influence with, and presence in, vitamin and supplement shops to rile the public. The NHA set a goal of raising $500,000 and generating one million letters to Congress within six months, and Gerald Kessler—who was so committed to the issue that he hired someone else to temporarily run his business—began personally lobbying members of Congress on the industry's behalf.

Deliberately ignoring the fact that the FDA's proposed rules would only regulate claims on supplement *labels*, not the products themselves, the NHA focused on making America's ten thousand or so health food stores into "Political Action Centers," designed to engage store employees and any consumer who walked through the door. Among other efforts, it sent a "Health Freedom Kit" to every store, encouraged the employees to talk about supplement regulation with every customer, set up letter-writing stations in stores, offered discounts to customers who participated (one San Francisco supplement chain offered customers a 20 percent discount "if you make your voice be heard"), and encouraged employees themselves to send a weekly letter to their congressmen. To make this letter writing easier, the NHA got the industry publication *Health Store News* to include preprinted letters in its spring issue that owners could sign and send to twenty-four different senators and representatives, all with the brilliant, Proxmire-inspired claim that the company sending the letter "Supports Freedom of Choice Regarding Natural Health Alternatives." As for customers, they were encouraged to participate in misleading letter-writing campaigns with slogans like "Write to Congress today or kiss your vitamins goodbye!" (which deliberately used the word "vitamin" to refer to *all* dietary supplements) and were subjected to planned "blackout days" on which supplement stores shrouded their products in black crepe and refused to sell them, presumably to demonstrate what would happen if the FDA got its way.

Donna Porter, a specialist in nutrition and food safety at the Congressional Research Service in the Library of Congress, was responsible for answering the hundreds of questions from Congress members that these letters provoked, particularly in regard to the erroneous claim that the FDA wanted to regulate all dietary supplements like prescription drugs. As she recalled, "When I pointed out to some industry lobbyists that what they were telling people to say in these letters wasn't true, they just shrugged and said, 'It works.'"

And indeed, it did work—as Porter described it, "The response was overwhelming." Countless individual customers began to take up the cause, sometimes quite creatively. A New Mexico woman appeared on television at a local protest wearing a Statue of Liberty costume, onto which she had pinned several herbal tea bags, proclaiming to the *Washington Post* that "what the statue really means for me is freedom of choice." (The tea bags were never fully explained.) As a result of the industry-led campaign, an estimated two *million* letters were sent to Congress; according to a 2000 article in *HerbalGram*, "No other law has ever received as much direct grassroots advocacy."

The FDA continued to insist that it didn't want to stop supplements from being sold or regulate them like prescription drugs; it just wanted to prevent companies from making unsubstantiated health claims on the labels. But to the industry and its supporters, the two were the same. "If the regulations go into effect," said Scott Bass, an industry lawyer, "the products will be taken off the market because the manufacturers won't take the health-claim labelling off. They are the lifeblood of the industry."

The result, as columnist Al Kamen summarized it in the *Washington Post*, was a battle that was "wacky even by Hill standards." As he explained it, "The enormous lobbying effort against the labelling law, which swept health food stores and generated tidal waves of calls, was directed at an outcome— loss of access to such remedies without a prescription—that the law never contained."

However, as Gerald Kessler had correctly predicted, the public reacted more strongly to the Nutritional Health Alliance's marketing blitz than it did to the FDA's actual proposal, and people seemed to unquestioningly equate "vitamins" with the general category of dietary supplements. Distrust of the FDA continued to grow. Then, in the spring of 1992, the FDA did something that massively hurt its own cause.

It started when the agency obtained a warrant to do a criminal search of an alternative medicine clinic in Washington State that the FDA claimed was "receiving, using, and dispensing several unapproved and misbranded foreign-manufactured injectable drug products." Previously, FDA investigators had found that the clinic was selling L-tryptophan, which at that point the agency had banned, and its owner was illegally manufacturing injectable high-dose vitamins in vials that had mold on them. When the owner learned that the FDA was investigating his clinic, he posted a sign on its door stating "No employee, agent or inspector of the FDA shall be permitted on these premises" and refused to let inspectors inside.

Given the owner's previous defiance, when FDA agents arrived to inspect the clinic on May 4, 1992, they were accompanied by local sheriffs. Unfortunately, those sheriffs—who'd been told the FDA was there to investigate "illegal drugs" (which was technically true, since the owner of the clinic was violating the law by manufacturing remedies himself)—assumed that the drugs in question must be heroin or cocaine, and that the confrontation might be violent. So when the owner of the clinic still refused to let the inspectors in, the sheriffs broke down the door, and one deputy drew his gun. According to the sheriff's office, it was never pointed at anyone and was quickly put away once the officer realized that the situation wasn't dangerous.

But that's not the story that went public. Instead, the raid was described by the PR Newswire as follows: "Fifteen Food and Drug Administration agents in black flak jackets, with guns drawn, backed by a contingent of armed King County police, broke down the door and stormed into the Tahoma Clinic of Dr. Jonathan Wright. In a scene that resembled a television drug bust, agents shouted at bewildered clinic employees and patients, 'Drop everything and put up your hands.'"

This description was quite exaggerated, given that the FDA inspectors did not have guns, and had obtained a legal warrant to search the clinic, with which the owner and employees had refused to comply. Nonetheless, the raid quickly became a public relations nightmare for the FDA. Someone had made a videotape of the raid; the health clinic's owner quickly made copies and began selling them to the public. An editorial in the *Seattle Post-Intelligencer* demanded an explanation for the raid's "Gestapo-like" tactics, and Senator Hatch introduced the Health Freedom Act—a law, never passed, that would have prevented the FDA from using supplements' health claims to determine whether they should be regulated as drugs.

Then in August 1992 (by which point there'd been plenty of time to correct the mistakes in the initial description of the incident) the *New York Times* ran an error-laden front-page article that still described armed FDA agents "dressed in bulletproof vests, bursting into the clinic and commanding clinic employees to freeze" before seizing more than $100,000 worth of medicine, office supplies, and equipment. The author also wrote that "last year, the [FDA] proposed regulations for the labelling law that would classify vitamins and minerals as drugs if dosages exceeded the daily recommended allowances; restrict or prevent the sales of most medicinal herbs like camomile; prevent unsubstantiated health claims for most dietary supplements, and lower the recommended vitamin-intake levels for various age groups." In reality, other than preventing unsubstantiated health claims, the rules would have done none of these things.

Indeed, the author of the explosive *New York Times* article had got so many of her facts wrong that the following Sunday's paper contained an eighteen-paragraph correction on the *front page*, including the clarification that the FDA officers were not actually armed, and an admission that "the article erred in saying that the proposed regulations would classify vitamins and minerals as drugs . . . and would restrict or prevent the sale of most medicinal herbs."

But the damage to the FDA's public image had been done, and the word "vitamin" continued to be used as a stand-in for the larger category of dietary supplements. Some two thousand letters were faxed to President Bush, and hundreds more were sent to the FDA. Celebrities began to rally against the FDA (the public should "start screaming at Congress and the White House not to let the FDA take our vitamins away," actor Sissy Spacek told the *New York Times*) and industry spokespeople denounced the FDA for overreacting.

"For God's sake, we're talking about vitamin C, B12 injections and Sleepy-time tea," said the executive director of a new organization called Citizens for Health, which was one of more than a dozen industry-sponsored "consumer

advocacy" groups formed in response to the raid.[3] By October 1992, Repre
sentative Waxman was forced to withdraw his proposed bill giving greater
enforcement authority to the FDA. Later, in August 1993, a San
Francisco-based supplement manufacturer created a sixty-second television
advertisement in which a gun-wielding SWAT team with night-vision goggles
raids Mel Gibson's medicine cabinet to get his vitamin C, with the (untrue)
claim that "the federal government is actually considering classifying most
vitamins as drugs," and an encouragement from Gibson to "call the U.S. Senate and tell them you want to take your vitamins in peace."

As the enactment dates for the new rules on *food* health claims drew near,
the supplement industry managed to get a moratorium on the regulation of
supplement health claims. And as *that* deadline approached, both sides
stepped up their fight for public opinion. In an attempt to widen its regulatory
net, the FDA expanded its definition of "supplements" to include products
like herbs and enzymes. It also stressed the issue of safety, with Commissioner David Kessler telling the *New York Times* that while "[t]he dietary
supplement industry is pushing hard for deregulation of their products,
[t]here are no assurances that these products are appropriately manufactured,
that what's on the label is actually in the bottle, that they bear adequate directions for use to insure safety or that basic safety data has been collected or
reviewed."

On the other side, Gerald Kessler continued his assault on behalf of the
supplement industry. But with individual politicians now receiving angry letters from hundreds of their constituents about their freedom to take
vitamins—and with the threat of Waxman's proposed bill out of the way—he
expanded his goal. Now he didn't just want to block the FDA's proposed
health-claim rules for supplements; he wanted to pass a supplement bill that
the industry could enthusiastically *support,* something that would free the
industry from the risk of FDA intervention once and for all. Several Hatch
aides and industry representatives got to work crafting such a bill, which Gerald Kessler christened the Dietary Supplement Health and Education Act.
Senator Hatch and Representative Bill Richardson (Democrat from New
Mexico, another supplement-heavy state) introduced the bill, and Tony
Podesta—one of the most influential Democratic lobbyists in Washington,

---

[3] Sleepytime tea would not have been affected.

whom Hatch and Richardson had convinced Gerald Kessler to hire—finagled hearings on it in both the Senate and the House. The bill had additional strong support from Senator Tom Harkin (Democrat from Iowa), who was convinced that bee pollen had cured his allergies.

At the House hearing, which began on July 29, 1993, arguments were passionate on both sides. Hatch claimed that dietary supplements had "been safely used for centuries" (despite the fact that there is often a big difference between, say, adding a pinch of a herb to a tea and ingesting a concentrated extract of that same plant) and called upon an AIDS patient who testified that patients' lives depended on access to them.

On the other side, FDA commissioner David Kessler made the point that the FDA's primary concern was not the many vitamin products and other dietary supplements that were safe; it was the fringes that were the issue. "When supplements are really drugs in disguise, promoted to treat serious diseases, we have a problem," he said.

He also emphasized the point that just because something is "natural" does not mean that it is safe. (In fact, the government still has not established a definition for the term "natural," despite years of trying.) "Think about it," said Kessler. "Half our prescription drugs are derived from plants, and no one doubts for a minute that drugs can have toxic effects. That is why we insist on rigorous testing to separate out those with unacceptable toxicity. We must not assume that all risk disappears when plants are sold as dietary supplements for therapeutic purposes."

To make his point about the issue of unsubstantiated health claims, he told his audience about an experiment run by the FDA in which FDA officials had done 129 informal surveys with salespeople at health food stores, asking the employees to recommend products that could treat serious conditions like cancer or infections. Ninety-three percent of the salespeople had complied. Then he had assistants load the witness table in the House Commerce Committee hearing room with hundreds of bottles of vitamins and other supplements that claimed to help everything from cancer to broken bones.

"We are back at the turn of the century," Kessler told his audience, "when snake oil salesmen could hawk their potions of promises that couldn't be kept."

In the midst of this maelstrom of exaggerated health claims, proposed

FDA rules, industry-generated outrage, and political manoeuvering, bets were on as to who would ultimately win the fight. The *New York Times* had described one far-fetched resolution in June 1993, before the House and Senate hearings: "The most militant industry members believe the only solution to their battle with the agency is in congressional legislation that would permit the industry, not the agency, to decide whether its products are safe and its labelling truthful," it said, referring to the bill that Gerald Kessler had helped create. "Under such self-policing, supplements would be considered safe unless the F.D.A. could prove otherwise." The idea seemed so outlandish that the article quickly moved on to describe proposals from "a more moderate faction." But, as it turned out, the militants would win.

My research of dietary supplement regulation happened to coincide with shoulder surgery. Since my recovery took several months, I became friendly with my physical therapist, and every week during our sessions, I'd update him on my latest supplement research. (Sometimes I used my newfound knowledge to get back at him for particularly painful exercises, like when I gleefully printed out articles from *Consumer Reports* showing that samples of one of his favourite protein powders had high levels of arsenic and lead.) One week, as he was helping me through an exercise for a rotator cuff muscle, he suggested that I include details on what percentage of supplements currently on the market had been approved by the FDA.

"Actually, none of them are," I told him as I sat on a yoga ball holding weights in both hands.

He looked at me suspiciously. "What do you mean?"

"There's no FDA approval process for supplements," I said, bouncing slightly. "And you don't need to test them for safety or efficacy, either."

Had I not spoken with my sea buckthorn saleswoman, I would have made the same assumption—namely, that there must be some government watchdog checking to make sure that supplements are safe. My therapist and I wouldn't have been alone: a 2002 Harris poll found that more than half of Americans believed that dietary supplements were approved by a government agency, and about two-thirds thought that the government required dietary supplement labels to include warnings about potential side effects.

Neither of those assumptions is true, because—despite the fact that it was opposed by Consumers Union, the Center for Science in the Public Interest, the National Organization of Rare Disorders, the American Association of Retired Persons, the American Cancer Society, the American Heart Association, the American Nurses Association, and the American College of Physicians—the industry-inspired Dietary Supplement Health and Education Act (DSHEA) was passed. Cosponsored by sixty-five senators, it was signed into law by President Clinton on October 25, 1994.

Reading from a statement (whose first draft had been penned by the supplement industry lobbyist Tony Podesta), President Clinton praised the legislation, commenting that "[a]fter several years of intense efforts, manufacturers, experts in nutrition, and legislators, acting in a conscientious alliance with consumers at the grassroots level, have moved successfully to bring common sense to the treatment of dietary supplements under regulation and law."

But when I asked former FDA commissioner David Kessler about his experience with the creation of DSHEA, which was technically an amendment to the 1938 Food, Drug, and Cosmetic Act, his perspective was quite different. "Is this the level of regulation that the people want?" he asked. "Or is this what the *industry* wanted? That's the question I still have today."

Either way, it's hard to overstate the impact that DSHEA—which is pronounced "D'Shea," as if Irish—has had on America's supplement market. Its first directive was to broaden the legal definition of the "dietary ingredients" in supplements (the technical term for the ingredients in dietary supplements) beyond vitamins and minerals to include less studied substances like herbs and botanicals, amino acids, enzymes, metabolites, organ tissues, and glandulars.[4] The FDA had suggested something similar in its proposed regulations, but whereas it had done so to expand its regulatory purview, DSHEA's goal was the opposite: to increase the number of products whose regulation the act would legally loosen.

In addition, DSHEA allowed supplements to take the form of liquids, powders, gelcaps, and foodlike items such as teas and bars, creating what Hurley describes as a "hermaphroditic category" between foods and drugs. It is the reason that, with the proper labelling, the energy drink Monster Energy

---

[4] According to DSHEA, a dietary supplement is "a product taken by mouth that contains a 'dietary ingredient' intended to supplement the diet."

could be sold as a dietary supplement instead of a food. But while surprising, this expanded definition of a "dietary supplement" was not the most consequential aspect of the new law. That honour was reserved for the dramatic differentiation it made between what it takes to bring a drug to market and what it takes to sell a supplement in the United States.

Over the past century, American pharmaceuticals have become the most respected and trusted in the world, thanks to the fact that every prescription and nonprescription drug must earn preapproval for its safety and efficacy from the FDA before it can be sold. It's a long process. As journalist Philip Hilts writes,

> Catastrophes like the one caused by Elixir Sulfanilamide have made it routine to check first whether the drug is an outright poison. Those tests are conducted on cells, then on two or more animal species. A drug's chemistry is also considered—can the drug be made reliably, its components rendered stable so it does not deteriorate on the shelf? Exposing animals to a drug can show whether it interferes with the normal chemistry of organs, whether it breaks down into other chemicals, and whether those "metabolites" are hazardous. Will it get into the blood effectively, or is it quickly flushed from the body? If not enough is absorbed, very large doses may be needed to cause the intended effect, and large doses in turn are often toxic. Then there are the questions about how the drug affects behavior. Do the animals become agitated when taking it, or unusually sleepy? Do they go off their feed or lose weight?

Answering these questions is neither easy nor cheap. And they're only the beginning: if a drug succeeds in the animal trials described above, it moves to human trials, beginning with a few dozen people (phase 1), then a few hundred (phase 2), and finally a larger group of several thousand (phase 3). The FDA also inspects the company's manufacturing facilities, mandates that all drug packaging include a detailed patient insert explaining the product's mechanisms and potential benefits and risks, and requires companies to keep track of and report adverse events once the drug is on the market. As a result, the amount of documentation required for a drug approval is astounding: just the data on safety can total more than ten thousand pages. And the amount of work required means that the number of "new molecular entities" approved each year—that's the technical term for what the public would just call drugs—is relatively low: between 2001 and 2010, it ranged from seventeen to thirty-six approvals per year. (The error rate is also low: the percentage of drugs approved between 1994 and 2004 that were withdrawn after approval due to unforeseen side effects and safety issues was 2.3 percent.)

These requirements for drug approval have undoubtedly kept dangerous and ineffective drugs off the market. But they're also time-consuming and costly. According to a 2013 analysis by *Forbes* magazine, some 95 percent of experimental medicines—that's nineteen out of twenty!—fail to meet the FDA's standards for safety and effectiveness for humans. Those that do succeed take roughly twelve to thirteen years to develop and gain approval. As for the financial cost, according to the *Forbes* analysis, companies that launched more than four drugs between 2003 and 2013 spent a median of $5.2 *billion* per drug on research and development, since for every drug that got approval, they also had to pay research and development costs for the nineteen or so that failed.

Thanks to DSHEA, none of these requirements applies to supplements—not the research, not the testing, and certainly not the cost.

To start with, DSHEA automatically grandfathered in all dietary ingredients that were on the market as of October 15, 1994, regardless of whether they'd ever been officially studied (which many botanicals, herbals, amino acids, and other non-vitamins and mineral products had not been).

DSHEA also loosened requirements for *new* dietary ingredients. Whereas new drugs are considered guilty until proved innocent (through the long FDA approval process), new dietary ingredients do not have to be preapproved. Instead, DSHEA dictates that companies' only requirement is to send the FDA information that indicates why they believe the new ingredient will be safe. There are no specific requirements for this information other than that it be submitted at least seventy-five days before the product goes to market; as the FDA website explains to would-be manufacturers, "You are responsible for determining what information provides the basis for your conclusion." (As of late 2014, there's draft guidance pending about what type of information should be submitted, but it hasn't been finalized. Also, FDA guidance is not binding.) And while DSHEA makes it clear that supplements are to be regulated more like foods than drugs, its requirements for new dietary ingredients are actually *less* strict than those for new food additives, such as new preservatives or colourings, which technically have to be approved by the FDA before being sold (though there are many instances of companies not defining ingredients as "new" in order to avoid this requirement). To put this in more concrete terms, if you wanted to add a new dietary ingredient—say, ground-up hamster brains—to a drug or a food, you'd have to spend several years and millions of dollars, if not more, to prove that it is safe and effective. If you wanted to add it to a supplement, no proactive testing would be required.

And that brings us to perhaps the most shocking consequence of DSHEA: despite the fact that, thanks to the Proxmire amendment, dietary supplements can contain nearly any combination of dietary ingredients in any dosage, today's supplement manufacturers do not have to prove that their products are safe or effective before selling them. Instead, the burden of proof is on the FDA to demonstrate, at taxpayer expense, that supplement products are unsafe *after* the products are already on the market. Thanks to DSHEA and all the consumers, industry representatives, and politicians who supported it, America's supplements have largely been made exempt from nearly a century's worth of tighter regulation for food and drugs.

Those in favour of DSHEA argue that the FDA is able to pull products off the market if they can be shown to be unsafe—and it's true that the 2011 Food Safety Modernization Act recently granted the FDA the new authority to issue mandatory recalls of products it deems to have a "reasonable probability" of being unsafe. However, eliminating dangerous products is much easier said than done.

Indeed, it's so difficult to do so that the FDA has banned only *one* dietary ingredient since the 1994 passage of DSHEA: ephedra—or, more precisely, ephedrine alkaloids—a dangerous stimulant often obtained from the plant *ma huang*. Despite overwhelming evidence of its dangers (it's thought to have contributed to the deaths of more than a hundred people), it took more than a decade of legal struggles and the highly publicized ephedra-related death of Steve Bechler, a twenty-three-year-old pitcher for the Baltimore Orioles, before the FDA succeeded. (Even today you can still find products being sold online that contain forms of ephedra.) The whole experience was so difficult and so costly that, in the words of the Government Accountability Office in 2009, banning further dietary ingredients—even though there are several definitively known to be dangerous—"is not a very viable option."[5]

Despite all this, the preamble to DSHEA—which, again, was written with direct industry help—claimed that the law was necessary because consumers "should be empowered to make choices about preventive health care programs based on data from scientific studies of health benefits related to particular dietary supplements." The federal government, it continued, "should not take any actions to impose unreasonable regulatory barriers limiting or slowing the flow of safe products and accurate information to consumers."

In reality, however, the law has made it substantially harder for "empowered" consumers to make choices among "safe products" based on "accurate information" and "scientific studies"—because it eliminated the requirement that such studies be conducted to begin with. And speaking of information, whereas most important laws include substantial legislative histories—that is, official records of the bill-making process, which can include committee reports, analyses by legislative counsel, committee hearings and floor debates, and which can be useful in interpreting the lawmakers' intentions—the chief sponsors of DSHEA deliberately restricted its official legislative history to seven sentences.

While it can be difficult to quantify the precise effects of a particular law, in the case of DSHEA, several figures stand out. As you may recall, the tightened regulations in the amended 1938 Food, Drug, and Cosmetic Act greatly decreased the number of medications on the market—it was largely responsible for Smith Kline's aforementioned decision in the 1950s to discontinue all but 60 of its 15,000 therapeutic products. In contrast, when DSHEA was passed, America's supplement market contained about 4,000 dietary supplement products. Today, there are more than 85,000.

In his history of the FDA, *Protecting America's Health*, journalist Philip Hilts includes a chilling description of the state of America's drug market after the passage of the 1906 Pure Food and Drug Act. "The law still permitted, as

---

[5] Instead, the FDA attempts to warn the public against potential dangers by issuing dietary supplement "Consumer Updates," which you can sign up for on the FDA's website, and supplement "Safety Alerts and Advisories."

muckrakers vehemently pointed out, that the human population of the United States be guinea pigs for all experiments with medicinal drugs," writes Hilts. "No testing was required before the drugs were sold, and if the federal government was concerned about any drug, it had to *prove* in court that the drug was harmful to a substantial number of people before removing it from the market. If the danger was subtle, a cumulative effect over a long period, or if it affected, say, only one in a thousand people, the damage would likely continue indefinitely."

It's now more than a hundred years later. But, thanks to DSHEA, you could be forgiven for thinking that Hilts is describing the state of America's supplement regulation today.

The supplement industry readily acknowledges its dependence on DSHEA— as the Council for Responsible Nutrition, a major industry trade association, states in a Q&A about supplement regulation on its website, "If dietary supplements were regulated like drugs, there would likely be no dietary supplement industry."[6] But as dramatic as they were, DSHEA's effects were not limited to safety and efficacy.

As you may recall, the industry-led advocacy effort that led to DSHEA was originally a response to concerns that the FDA's proposed restrictions on supplement health claims might affect sales—a legitimate worry, given that few consumers would buy ground-up adrenal glands or a product called N.O.-XPLODE without at least some sense of what they were supposed to do. DSHEA resolved that issue as well—and, again, the industry won.

This victory came in the form of structure/function claims, carefully phrased statements that explain how a particular ingredient might alleviate a nutritional deficiency, improve the structure or function of a particular part of

---

[6] The quote continues by claiming that if supplements were regulated as drugs, "supplements would cost what drugs cost." That is highly debatable, since few people argue that dietary supplements should be regulated as *prescription* drugs. A better comparison might be over-the-counter drugs, which include not just obvious things like Tylenol but "cosmetic" products with druglike attributes, such as toothpaste with fluoride, deodorants with antiperspirant, and anti-dandruff shampoo, all of which are technically considered drugs by the FDA (they all have Drug Facts panels on their boxes) but are certainly not prohibitively expensive.

the human body, or simply promote well-being. More practically speaking, they're the bold-print claims you see nearly every time you pick up a supplement box—the ones that say a product "promotes healthy cholesterol levels" or "supports healthy intestinal flora" or "helps maintain cardiovascular function."

Like high school guidance counsellors, structure/function statements always use very positive and encouraging verbs, and to avoid qualifying as disease claims, they are not allowed to contain the words "treat," "mitigate," "cure," or "prevent." Instead, the language used is reminiscent of the "I" statements endorsed by couples therapists to help people express their opinions without making their partners feel attacked. Whereas a therapist might propose replacing "You never listen to me" with "I feel like I'm not being heard," a supplement maker will avoid the assertive phrasing of "prevents urinary tract infections" in favour of the gentler "improves urinary health."

Not surprisingly, structure/function claims are a political creation, conceived on October 7, 1994, when Gerald Kessler and his lobbyists were engaged in a last-minute effort to get the House of Representatives to release DSHEA from its subcommittee and put it up for a vote before Congress adjourned. Originally, Kessler had wanted supplement labels to be allowed to claim the product would treat, cure, or prevent a disease. When FDA officials and lawmakers refused—that is, after all, the definition of a *drug*—the politicians created structure/function claims as a compromise.

Structure/function claims do not have to be preapproved; a manufacturer must simply alert the FDA to the structure/function claim within thirty days *after* marketing the product and have some sort of substantiation for the claim on hand that hypothetically could be used to demonstrate that it is "truthful and not misleading." Perhaps these loose standards are part of the reason why, in a late-night concession to the FDA, the supplement industry agreed to one other requirement: when a supplement uses a structure/function claim, its label must include a statement, set off by an asterisk, that will be familiar to anyone who has bothered to read the box: "These statements have not been evaluated by the Food and Drug Administration. This product is not intended to diagnose, treat, cure, or prevent any disease." The compromise enabled the bill to be released from committee and passed.

But while structure/function claims will hold up in court, they're also, from a commonsense perspective, fundamentally absurd. Why would you take a supplement, after all, if you didn't think it could treat, cure, or prevent something? Are there really people who would buy a product that "improves regularity" without hoping it might also treat constipation? Structure/function statements—which are now nearly ubiquitous on dietary supplement labels— allow companies to make far more assertions than they're permitted to in the more strictly regulated category of food health claims, and few consumers know the difference. Nor do consumers realize the circular process some companies use to create them. As an employee in the supplement industry once explained it to David Kessler, "We use focus groups to decide what to say in claims. We hear what people want, and then we put that on the label."

Today, more than twenty years after DSHEA's enactment, most consumers are not aware of its existence, let alone its effects on our lives. But even if we did know about it, we seem unlikely to care. Just as we often use the word "vitamin" to apply to any nonpharmaceutical pill, we extend vitamins' known and studied safety profiles to all other supplements. As the word's creator, Casimir Funk, would be proud to observe, its appeal and power have only grown. As a result, while most people would hesitate before experimenting with random combinations of prescription drugs, we feel no qualms about doing the same with dietary supplements.

Consider the example of a product called Natural Curves, available for about $24.99 a bottle online and in stores. Its packaging features a large photograph of a woman's cleavage, but in case that does not give you an adequate sense of its intended purpose, here's its description on GNC's website:

> 100% Natural
> BREAST ENHANCEMENT
> Natural Bust Enhancement
> Balanced Formula for Maximum Results
> 100% Natural
> DIETARY SUPPLEMENT
> 60 TABLETS

30 DAY SUPPLY

Let Natural Curves™ help give you the self-confidence to look
and feel your best, whether it's a day at the beach, a night on the
town or a quiet night in front of the fireplace. Get the curves
you've always wanted naturally . . . with Natural Curves.™
IMAGINE THE POSSIBILITIES.

In addition to its implied promise that it will provide a "natural" boost to a
phase of development that usually ends with high school, the label lists its
ingredients as a proprietary blend of herbs and botanicals that includes saw
palmetto berries, blessed thistle leaf, wild yam root, mother's wort leaf, and—
somewhat incongruously, given the box's main graphic—chasteberry root.
According to its label, Natural Curves' "natural" effects are due to its "natural
blend of isoflavones and other key herbal extracts" that "'balance' the levels
of key 'breast-sensitive' hormones in a woman's body and thereby helps to
maximize the growth of breast tissue."

When a search on PubMed did not reveal any randomized, controlled
studies demonstrating how exactly this enhancement was supposed to happen, I turned to an alternative: customer reviews. The supplement had earned
3.5 stars out of 5, though its forty-nine reviewers, including one woman who
aspired to a 38DDD "or a little bigger," showed a surprising lack of correlation
between the ratings they gave and their reported results.

"I have bad acne . . . [and] when I started taking Natural Curves, it made
me break out a LOT," wrote one woman—who proceeded to give the product
four stars and said she would recommend it to friends. "So I stopped taking it
after 3 weeks and the results quickly disappeared. :("

Many women were enthusiastic, even in the face of adversity:

I had heard that these particular herbs can negatively affect the hormonal
balance of a younger person but have had no problems so far. I just bought
my second bottle today!

The downside is the acne. It was like I was a preteen all over again. The
acne became overwhelming so I stopped taking half way thru a bottle.
Within a week or so, I noticed my plump little breast [*sic*] return to little

deflated balloons. I immediately re-started the pills. I'll have to find another way to fight the acne.

These user comments highlight one of our strangest assumptions about supplements: that something can be simultaneously innocuous *and* miraculous. It's an assumption inspired by vitamins, which, at the levels found in food, can indeed be harmless and lifesaving at the same time. But it's also delusional.

In the case of Natural Curves, if the reviewers are correct and it's causing acne (in addition to whatever it's done with their breasts), then the label is accurate: it is indeed having hormonal effects on their bodies. But *what* effects, exactly?[7] (Imagine the possibilities!) The scariest supplements are not the ones that are totally ineffective; they're the ones that actually *do* stuff, since—thanks to DSHEA—manufacturers are not required to study exactly what they're affecting, how they work, or what their long-term effects may be (let alone how they interact with other ingredients, foods, or drugs). Most people would not take a pill offered to them by a stranger in a club. So why is there a market for Natural Curves?

Cleavage-enhancers aside, there are many dietary ingredients in supplements that have been evaluated for safety (though usually not by their manufacturers), and our reaction to this research highlights something arguably even stranger: we continue to believe in dietary supplements' inherent safety despite the fact that many supplement ingredients that are allowed to be sold in the United States have been definitively proven to have both short- and long-term health risks and have been banned in other countries. *Aristolochia,*

---

[7] The acne may be related to the chasteberry, according to the National Center for Complementary and Alternative Medicine. While NCCAM says nothing about chasteberry and cup size, it cites research finding that chasteberry can affect "certain hormone levels" and is not recommended for pregnant women, women on birth control or with hormone-sensitive conditions like breast cancer; it also may affect the dopamine system in the brain and therefore shouldn't be used by people taking dopamine-related medications, like some antipsychotic drugs or Parkinson's disease medications. If this sounds worrisome, remember that chasteberry is only *one* of Natural Curves' ingredients.

which can be an ingredient (or substitute ingredient) in Chinese herbal products, is carcinogenic, can cause kidney failure, and has been banned in at least
seven European countries, as well as Japan, Venezuela, and Egypt. Comfrey,
chaparral, germander, and kava can cause severe liver damage. Kava has been
banned in Canada, Germany, Singapore, South Africa, and Switzerland; germander is banned in France and Germany.

Even such generally safe substances as garlic, ginkgo biloba, ginseng, and
vitamin E can all cause blood thinning, which may lead to life-threatening
complications during surgery—which is one of the many reasons you should
tell your doctor about all supplements that you're taking. Aloe vera extract
contains strong laxative compounds, and when taken orally, has been shown
to cause cancerous tumours in rats. A quick look through some of the side
effects mentioned on the NIH National Center for Complementary and
Alternative Medicine's "Herbs at a Glance" page makes it immediately clear
that, to reiterate the message of former FDA commissioner David Kessler,
just because something is considered natural doesn't mean that it's safe.

Thankfully, most supplements are usually not harmful, at least not in the
short term. As Tod Cooperman, the president of a supplement testing company called ConsumerLab.com, put it to me, "In the vast majority of cases,
people are just throwing their money out the window or hurting themselves
over a long period of time." But as the last part of his comment suggests, there
are still reasons for concern over what's in some of these products—especially
given that in the most recent data available from the Centers for Disease Control and Prevention, more than 50 percent of American adults reported using
some sort of dietary supplement.

And even vitamins come with caveats, though they're rarely divulged
unless it's required. For example, prescription prenatal vitamins—which are
standard multivitamins with some extra folic acid and iron—come with a
"patient prescription information" sheet that's mandatory for all drugs. It
includes a long list of possible side effects and interactions, including how the
tablets could potentially reduce the absorption of other drugs like common
antibiotics and thyroid medications, or mask the signs of a vitamin B12 deficiency. The sheet also lists possible side effects (diarrhoea, constipation,
upset stomach), tells you what to do if you have signs of a serious allergic reaction, and suggests that you tell your doctor or pharmacist about any other
medications or nutritional supplements you are taking at the same time in

case there might be additional interactions. The same formulation as an over-the-counter supplement would not require this information or warning.

One of the best-known examples of the dangers of our unquestioning faith in the safety of supplements is St. John's wort, which is an extremely popular "natural" treatment for depression (and one of the most-studied botanical dietary supplements in the world). As former FDA commissioner Jane E. Henney explained in the *Journal of the American Medical Association*, St. John's wort—which you could go out and buy right now in your grocery store's vitamin aisle—"interacts with many drugs that are used to treat heart disease, depression, seizures, certain cancers, as well as drugs that prevent transplant rejection and pregnancy."[8] Putting aside the irony of a herbal depression treatment that might interfere with your prescription depression medication and create the potential for an unplanned pregnancy to boot, St. John's wort could interact with your prescription drugs in a potentially life-threatening way (for example, by triggering an organ transplant rejection). And yet it's not required to have any sort of warning label.

Seeking the supplement industry's perspective on the interaction issue—particularly in regard to St. John's wort, since it's so well established—I called Steve Mister, president and CEO of the Council for Responsible Nutrition. He told me that, as far as CRN was concerned, the responsibility to warn consumers of possible side effects should not fall on the supplement companies. Instead, he compared St. John's wort to grapefruit, which interacts with many prescription drugs but, as food, is not required to have a warning label—and said that the responsibility should fall on the doctor prescribing the medication.[9]

---

[8] The primary compound in St. John's wort that's responsible for these interactions (as well as any antidepressant effects) is hyperforin. Among other actions, hyperforin increases the production in the liver and small intestine of an enzyme called CYP3A4, which is responsible for metabolizing (i.e., regulating the breakdown of) upward of 50 percent of conventional medications, says Bill Gurley, PhD, professor of pharmaceutical sciences at the University of Arkansas for Medical Sciences College of Pharmacy. The more CYP3A4 your body produces, the more efficient your body will be at breaking down the drug, and the smaller the amount of the active form that will get into circulation. "St. John's Wort renders most drugs ineffective," warns Gurley—who also points out that, thanks to the herb's ability to render birth control ineffective, "there are a lot of miracle babies associated with the use of St. John's wort."

[9] Grapefruit contains chemicals that affect the bioavailability of many drugs, including

"The impetus is on the healthcare provider who's prescribing the prescription drug to tell you, 'I'm going to give you this prescription, but while you're taking it don't drink grapefruit juice, stay away from tomato juice . . . these things will potentially interfere with the ability of the medicine to work,'" Mister told me. "The same should be true with supplements. Healthcare providers should ask you, 'What supplements are you taking?' and if there's a potential for interaction that's going to decrease the efficacy of the drug they should say, 'You know what, you ought to stop taking that,' just like they'd say stop drinking grapefruit juice."

At first, Mister's point might seem valid: St. John's wort is common enough that you'd think most doctors or pharmacists would be aware of its potential side effects. But that's putting a lot of faith in (and responsibility on) doctors, most of whom receive extremely little training in nutrition and dietary supplements to begin with, and who already have to remember to consider potential interactions and side effects of thousands of prescription drugs.

What's more, a 2007 paper published in the *Archives of Internal Medicine* suggested that many doctors might not recognize the need to put supplements under special scrutiny, since they are just as clueless about supplement regulation as their patients. Of 335 residents and attending physicians at fifteen different internal medicine residency programs, the study found that "one third of physicians were unaware that dietary supplements did not require FDA approval or submission of safety and efficacy data before being marketed," a similar percentage believed there were regulations to ensure supplement quality (at the time there were not), and "most physicians" were unaware that serious adverse events due to the use of supplements should be reported to the FDA.

As if that's not problematic enough, St. John's wort, which is identifiable, well studied, and usually goes by only one name, is not an appropriate stand-in for the thousands of other supplements on the market. Not only do many people conceal their supplement use to their doctors, but many less conventional supplements contain proprietary (read: undisclosed) blends of ingredients

---

antianxiety medications, antihistamines, and statins, and these effects can linger for hours—if not days—after you eat it. Unlike St. John's wort, grapefruit *inhibits* the activity of CYP3A4 in the intestines, which means that the body breaks down and excretes less of the drug than it would otherwise, which in turn means that more of the active form of the drug ends up sticking around in your body.

that may not have been studied for possible interactions with prescription drugs on their own, let alone in combination—and whose side effects doctors therefore have no way to predict. This is potentially so problematic that Cooperman, president of ConsumerLab.com, suggests avoiding any product that contains a "proprietary" formula, since there's no way to know what it contains.

Even if these substances had been studied, there's no guarantee that the doctor would be able to find them in a database or index, thanks to the fact that many dietary ingredients—particularly herbals and botanicals—can be referred to by more than one name or spelling. It's also common in traditional Chinese medicine to substitute herbs without revealing the substitution.

Conversely, the terms used for many Chinese herbs can also refer to more than one plant; according to Veterinarywatch.com, the term *jin qian cao* refers to "five different plant species, in five genera, in five botanically unrelated families. If a bottle or prescription simply lists the Pinyin [transliterated] name, there is no way to determine which plant—or part of the plant—is involved." There's also more than one dialect and more than one way to transliterate Chinese, leading to different English spellings for the same sounds. And those are just discrepancies within one language. Go to an Ayurvedic healer, and you'll have to know the plants' names in Sanskrit.

Even if you could correctly identify what's *supposed* to be in the bottle, there would be no guarantee that its contents are what its label claims. In 2013, Canadian researchers using DNA testing found that of the forty-four products they tested, a full third did not contain *any* amount of the substances listed on their labels. As the *New York Times* summarized it, "Many pills labelled as healing herbs are little more than powdered rice or weeds."

So-called herbal preparations have also been found to contain substances that do not come from plants or are not listed on the label, and that the supplement manufacturer itself may not know are in its raw ingredients. In addition to genetic material from endangered species, DNA analysis of supposedly herbal ingredients has revealed substances including antelope and deer horn, donkey skin gelatine, earthworms, human placenta, bat faeces, cicada exoskeleton, wingless cockroach, bear gallbladder, charred human hair, toad skin secretion, and seal penis. As Edzard Ernst, director of complementary medicine at Peninsula Medical School in Exeter, England, told *Nature* in 2012, "Many of those traditional Chinese medicine supplements are such adventurous mixtures of multiple ingredients that, quite frankly, nothing surprises me about them."

But as disgusting as it may be, seal penis seems relatively innocent, given the other things being mixed into supplements. The Canadian research mentioned above found some echinacea supplements that contained a weed linked to rashes, nausea, and flatulence; a bottle of St. John's wort (which contained no St. John's wort) that showed the DNA signature of an Egyptian plant known for being a strong laxative; and ginkgo biloba supplements that contained unlabelled nuts, thereby putting allergic consumers' lives at risk.

Perhaps most shocking of all, there are many supplements that have been spiked illegally with prescription drugs. I first became aware of this issue when I heard James Neal-Kababick, the founder and director of Flora Research Laboratories, describe how he'd once opened up a capsule of a supposed Chinese sexual enhancement herb to test it for possible adulterants— only to see a piece of a Viagra tablet tumble out. "That one was pretty easy," he told me.[10]

Unfortunately, this type of adulteration is common. The practice of spiking dietary supplements with prescription drugs is so problematic that numerous Olympic athletes who take nothing but legal supplements have been disqualified from competition for testing positive on drug tests. The US Anti-Doping Agency has an entire web page, "Supplement 411," devoted to warning athletes about their risks. In 2010, FDA commissioner Margaret Hamburg wrote an open letter to the supplement industry stating that "FDA laboratory tests have revealed an alarming variety of undeclared active ingredients in products marketed as dietary supplements," including anticoagulants, anticonvulsants, HMG-CoA reductase inhibitors, phosphodiesterase type 5 inhibitors,

---

[10] What's more difficult, he said, is when suppliers design clandestine analogues that are similar to Viagra but structurally different—in other words, a substance that is chemically related to Viagra and produces its effects, but that doesn't have Viagra's chemical signature and thus can slip by undetected. "We're up to 70 analogues like that," said Neal-Kababick. Sometimes supplement makers are aware of these compounds' inclusion, but other times they're not: "A lot of people buy a product as a pre-mix, ready-to-encapsulate," he said. "So they get this material from China that's supposed to be a blend of, say, 10 herbal extracts that are formulated in a secret ancient way that exactly replicates the effects of Viagra. They get excited, they spend $50,000 to buy a shipment, they send it to us, and we find an analogue of Viagra."

nonsteroidal anti-inflammatory drugs, and beta-blockers, as well as controlled substances like anabolic steroids and banned drugs like fenfluramine.[11]

Recently implemented good manufacturing practices (GMPs), which became fully mandatory in 2010, attempt to address this problem by requiring supplement companies to provide FDA inspectors with paperwork demonstrating that their products contain the ingredients that they claim. Unfortunately, however, these GMPs do not address the issue of "dry labs," contract laboratories hired by supplement manufacturers to do quality assurance tests on products or ingredients—but that don't actually do any testing. Instead, they just provide manufacturers with clean paperwork. Some supplement companies deliberately work with dry labs; others may be being duped themselves. Either way, since the FDA doesn't inspect contract labs, it's difficult for an FDA inspector to tell whether a manufacturer's paperwork is based on legitimate information.

Sometimes contaminants like arsenic, lead, and pesticides make their way into supplement products. Sometimes tablets or capsules themselves don't break down properly, and their contents are excreted before they can be absorbed. Sometimes the wrong species or part of a herb is used (like the leaves instead of the root). Sometimes research papers don't describe precisely which form or source of a herbal ingredient was used in the experiment, or supplement bottles don't reveal which ingredients they contain, making it difficult to ensure that the supplement you buy is the same formulation used in the trial that inspired you to take it.

Even when the ingredients are what's listed on the label—and no Viagra or seal penises are included—there's still no guarantee that they exist in the amounts or strengths that are claimed. Herbs' potencies can vary from one crop to the next (in fact, the levels of plants' chemicals, like human hormones, can fluctuate throughout the day, and often exist in different concentrations in different parts of the plant). Vitamins in particular are extremely sensitive to degradation, so manufacturers often include "overages" to make sure that

---

[11] As Hamburg herself notes, sexual enhancement, weight-loss, and bodybuilding supplements are notorious for adulteration (often with prescription drugs)—beware! And while this type of adulteration is not usually an issue for vitamins, it still does occasionally occur: one brand recalled some of its multivitamin products in June 2013 because they were contaminated with anabolic steroids.

vitamins will still contain at least 100 percent of the amount listed on the label up to their expiration dates. Most of the time these overages are harmless, but in certain cases—such as vitamin A—they can lead to supplements that contain overages high enough to be a concern, especially for children. And sometimes, there are just blatant formulation errors: in 2008, for example, more than two hundred people were poisoned when a multivitamin/multimineral product that said it contained 200 mcg of selenium per serving actually contained 40,800 mcg. In short, the list of potential contaminants and variables is long.

There have been several important regulatory developments since the passage of DSHEA. The Bioterrorism Act of 2002, which was strengthened by certain provisions in the 2011 Food Safety Modernization Act, mandated that supplement manufacturers register with the FDA (previously there had been no such requirement), which means that the FDA presumably at least knows the names and contact information of the companies that sell supplements in the United States—though as indicated by its website's FAQs, that list is not public. It also gave the FDA the aforementioned power of mandatory recall, though it can't be used till after the product is on the market. The 2006 Dietary Supplement and Nonprescription Drug Consumer Protection Act requires supplement makers to report to the FDA any "serious adverse events" related to their products. (For the first thirteen years of DSHEA, there were no reporting requirements for supplement-related adverse events at all.) The Federal Trade Commission, the agency that regulates supplement *advertising*—as opposed to labels and packaging, which are regulated by the FDA—has also become more aggressive in going after supplement companies making druglike claims.

And in 2007, the FDA finally issued the GMPs for supplements mentioned above. Written with direct help from the supplement industry ("We're in the cheering section on GMPs," said Steve Mister of the CRN, pointing out that they improve consumer confidence), they represent an attempt to ensure that products contain pure and unadulterated ingredients in the amounts their labels claim. According to the analytical chemist James Neal-Kababick, the quality of supplements on the market is better than it was before the GMPs were implemented.

But, Neal-Kababick continued, "that doesn't mean that there are not still issues," and significant loopholes—besides having no pre-market testing requirements for supplements' safety or efficacy—still remain. For example, a "serious adverse event" is defined as an "event that results in death, a life-threatening experience, inpatient hospitalization, a persistent or significant disability or incapacity, or a congenital anomaly or birth defect"—or one that requires a doctor's or a surgeon's intervention to prevent one of the above events from occurring. A trip to the emergency room, a night spent throwing up, a rash, an allergic reaction, or any other type of non-life-threatening reaction need not be reported to the FDA. These "nonserious adverse event" reports must be kept by manufacturers for six years, but don't have to be presented to the FDA inspectors unless requested.[12] Side effects can also take a long time to develop, which makes them difficult to link to a particular product, which in turn makes it less likely that companies will even receive adverse event reports about them to begin with.

The requirement that supplement manufacturers must register with the FDA is a step forward, but it's impossible to tell if all supplement manufacturers are complying with the requirement to register with the FDA, since the list only contains manufacturers who have proactively registered. Supplement manufacturers are also not required to tell the FDA what products they make, or what ingredients their supplements contain. This means that if a dietary supplement ingredient is found to be causing problems (as happened with L-tryptophan), there is no way for the FDA to identify and contact all of the ingredient's suppliers or sellers, because there is no database—let alone a searchable one—of who sells what.

As for the good manufacturing practices, not only do they not address safety or efficacy (they're focused on ensuring supplements contain the

---

[12] Before the reporting of serious events was made mandatory, Metabolife International, a leading manufacturer of the aforementioned stimulant known as ephedrine alkaloids (the only dietary ingredient the FDA has banned since the passage of DSHEA), failed to report to the FDA that it had received 14,684 complaints of adverse events related to its ephedra product, Metabolife 356, over the previous five years—including 18 heart attacks, 26 strokes, 43 seizures, and 5 deaths. Today, the FDA estimates that if manufacturers were required to submit *all* reports of adverse events related to supplements each year (including mild, moderate, and serious) the total would be more than 50,000. And even that might be an understatement. Medical statisticians typically assume that the ratio of reported to actual cases is between 1:10 and 1:100—and that's for foods and drugs, for which the cause and effect might be more immediately obvious.

ingredients they claim to contain, not whether those ingredients themselves
are effective or safe), but their enactment has revealed numerous problems
with quality control. When I spoke to the FDA's former director of Dietary
Supplement Programs, Daniel Fabricant, about two years after GMPs had
become mandatory for all manufacturers, he told me that at that point the
FDA had issued warning letters or taken some sort of legal, regulatory, or
administrative action against approximately 25 percent of the facilities that it
had inspected. This percentage was "very troubling," he said, especially since
the inspectors were focusing on "real basic evidence," like whether manufacturers were doing identity testing on their raw ingredients, or even had master
manufacturing records—that is, recipes for how they make their supplements.
(To put this in context, the rate of this type of violation among food manufacturers was closer to 3–6 percent, he explained.)

"We're really just asking the first layer of questions right now," said Fabricant. "What happens when we start drilling down in some of these areas? Given
that we're seeing in some cases people not being brilliant on the basics, we have
some concerns about whether they'll be adept on the more sophisticated level."

Unfortunately, two years after Fabricant's comments—and four years after
the GMP requirements went into effect for all supplement manufacturers—
the FDA still had many reasons for concern. According to Angela Pope, consumer safety officer in the Division of Dietary Supplements at FDA, the FDA
found serious GMP violations at 28 percent of the facilities it inspected in the
fiscal year that ended in September 2013. "All of the companies we've
inspected that have more than five hundred employees have had GMPs in
place," she said. But many smaller manufacturers do not. "For those that don't,
it's a scary, scary story," said Pope. "Very scary."

And who knows what the FDA is missing? In fiscal year 2013, the FDA
inspected about 600 domestic dietary supplement manufacturers (out of an
estimated total of between 1,600 and 2,800), which is up from 150 or so two
years earlier.[13] (The international numbers are even less concrete, but Fabricant has estimated that there may be as many as 300,000 foreign supplement
makers in total; in fiscal year 2013, the FDA inspected approximately 100.)

---

[13] These are mostly companies that take raw ingredients and premixes and put them into
  supplements; they don't actually produce synthetic vitamins or raw ingredients themselves.
  (As discussed earlier, bulk synthetic vitamins are nearly entirely produced abroad.)

The entire budget for the FDA's Division of Dietary Supplements was $10 million in 2004, and when the GMPs were issued in 2007, Congress did not allocate any extra money for the FDA to conduct inspections to enforce them. While the Division of Dietary Supplement Programs also relies on specialists elsewhere in the agency to carry out its duties, at the time when I spoke with Fabricant, the program itself had a total of twenty-four full-time employees to regulate an industry that in 2012 brought in more than $32 billion in US sales.

Any efforts to tighten supplement regulations continue to be met with strong opposition. The proposed Dietary Supplement and Awareness Act, for example, would require supplement manufacturers to register their products and ingredients with the FDA in addition to their names and contact information—thus making it easier for the FDA to take quick action in a public health emergency—but the bill had only three cosponsors and has been stalled in the House Subcommittee on Health since 2005. In 2010, Senator John McCain introduced a bill that would have realized many of the recommendations made in a 2009 report from the US Government Accountability Office titled *Dietary Supplements: FDA Should Take Further Actions to Improve Oversight and Consumer Understanding.* But the industry-generated opposition to the bill was so strong that McCain withdrew his support.

And despite the fact that some industry members believe that more stringent regulation would help business by increasing consumer confidence, the industry's official stance is that the supplement market is already more than adequately regulated, and that anyone who voices doubt or expresses criticism is getting the story wrong. Indeed, when I asked Steve Mister from the Council for Responsible Nutrition what he thought the media and the public misunderstood about America's supplement regulations, he laughed.

"Do you have three hours?" he asked. "One of the things we do here is play truth squad with the media when they report over and over again that dietary supplements are not regulated. The regulation for dietary supplements is quite comprehensive. . . . [Y]ou must list your ingredients, you must list them in a certain way, you must list what percentage of Daily Value there is, you must have your name and address on there so that people can contact you with adverse events.[14] The manufacturing is regulated, the formulations are

---

[14] Other than vitamins and minerals, extremely few dietary ingredients have RDAs or DVs to begin with.

regulated in some cases. Now, yes, you have a lot of flexibility . . . [but] there is a *lot* of regulation around these products. . . . Absolutely the regulation for supplements is comprehensive, it is robust, and it does give consumers adequate protection. Ample protection, actually."

Daniel Fabricant would not reveal his own thoughts on whether the fact that a company must reveal its name, address, and product ingredients is really proof of the adequacy of America's supplement regulation. But when I asked him the same question I'd posed to Mister—about what the public misunderstands—his answer was quite different. "I think a lot of people assume that because it looks like a pill, somehow the FDA has signed off on the safety of the product," he said. "We always caution people against making this assumption."

Given all these issues, it is extremely difficult—if not impossible—for consumers to identify high-quality dietary supplements in the store.

"It's really hard to say this company's good or this one's bad or this one's products are always reliable and this one's are shaky," said Neal-Kababick when I asked him for advice. "It depends not just on the company and where they are in that point of time, but on what's going on in the demand and supply of the market. When demand is high and supply's low, adulteration floods the market."

Some companies voluntarily submit their products for certification by third-party testing companies like NSF and the US Pharmacopeial Convention, whose USP Verified Dietary Supplement seal should not be confused with the general acronym USP, which supplement makers sometimes use on packaging or in brand names. Consumerlab.com is a particularly useful resource, as it's the only testing organization that pulls products randomly from store shelves. (Keep in mind, though, that these companies' tests are only designed to verify ingredients and, in some cases, whether the manufacturers are following GMPs. They do not evaluate safety or efficacy.)

The best advice about supplements, however, is less practical than it is philosophical. Before you pick up a box or bottle, ask yourself *why*, exactly, you're buying it. What do you think it will do? What evidence do you have for your beliefs? What are its known side effects and interactions? Is there a chance that it might do more harm than good?

In a broad sense, there are many reasons you might buy a supplement. Perhaps, as was true for me and my skin issues, you're frustrated by the uncertainties of modern medicine and long for a sense of control. Perhaps you distrust pharmaceutical companies or are angry at their profits. Perhaps you fear overmedicating. Perhaps you long for some sort of spiritual aspect to your health care, or put more trust in the safety of traditional or "natural" remedies than you do in prescription drugs. Perhaps you chafe against the idea of "big government." Perhaps you feel abandoned by your doctor, or believe that supplements are your only hope. As Representative Henry Waxman explained to journalist Dan Hurley, "When people are facing health insurance that they can't afford, and treatments for cancer that sometimes result in side effects that are just as awful as the cancer, they're ready for a cure—they're desperate for one. They would like to believe that there is some kind of conspiracy that's keeping them from knowing the true facts that could keep them healthy." Waxman, speaking specifically about the public's advocacy efforts on behalf of DSHEA, continued, "What they didn't understand, though, was that this view was manipulated by people who stood to make a lot of money, and they did make a lot of money—billions of dollars."

These profit margins become even crazier when you realize that in many cases, products don't actually contain the ingredients you think you're paying for. Many people turn to supplements to boycott "big pharma"—but what about "big supplements"?

In the 1993 congressional hearing about DSHEA, then commissioner Kessler summarized the issue in a way that is still relevant more than twenty years later. "Much is at stake for consumers," he said. "Believe me, I appreciate as a physician the appeal of a simple cure. Of course, we would all rather take some miracle pill than undergo more arduous and sometimes uncertain treatments. But, unfortunately, cures don't come packaged as neatly as we hope, and patients, who would forsake therapies that offer some real benefit for the siren song of empty promises, have a lot to lose."

By the time I'd begun to wonder about the philosophy of dietary supplements, I'd nearly forgotten about sea buckthorn, that coastal shrub whose promises of itchy skin relief had launched my journey into the netherworld of supplement regulation to begin with. When a new wave of skin irritation reminded me, I did some searching and managed to find the study about sea buckthorn oil that seemed the most likely to be the one mentioned on my supplement label. The study, which was a placebo-controlled, double-blind trial of forty-nine patients, concluded that while sea buckthorn *seed* oil didn't cause any significant improvement, sea buckthorn *pulp* oil did help the dermatitis.

That sounded great, except that the study found that paraffin oil—that is, the control—was also effective in treating atopic dermatitis. What's more, the study required participants to take five grams of sea buckthorn pulp oil a day, every day, for four months. Since my sea buckthorn supplement label didn't single out pulp oil as a separate ingredient, I didn't know exactly how many capsules I'd have to take to equal that dose—but at the very least it would be more than ten a day (and the regimen would be quite expensive, considering that a sixty-capsule bottle costs more than $30). Finally, the study also found that the increase in palmitoleic acid caused by the pulp oil might have been related to a higher level of LDL cholesterol—the supposedly bad kind.

My dermatologist had never heard of sea buckthorn oil; it's not listed in the National Center for Complementary and Alternative Medicine's database (which is a useful resource for herbs and botanicals), and when I looked it up on ConsumerLab.com, their conclusion was nearly as concise: "There are no well-established therapeutic uses of this herb."

If any of this disturbs you, you might be tempted to blame the FDA for not being more aggressive—that's what the supplement industry often does. But as we've seen, the agency has surprisingly little control over its supposed powers; in the words of Fabricant (who has since left the FDA to become chief executive officer of the Natural Products Association, the United States' largest trade organization in the so-called natural products industry), the role of the FDA is to "take the law from the books and maximize how we can protect public health." Those laws are created by Congress. And members of Congress are accountable to their constituents who vote for them. Yes, the industry's lobbyists are extremely powerful, but their salaries are paid out of money from consumers like you and me. Not only do we financially support the supplement industry by buying its products, but by responding to its "grassroots" advocacy campaigns, we do its work for it: every time the FDA has proposed tougher regulations, it is we the people who have ultimately forced it back in line.

So perhaps it's time for a bit of self-reflection. Are we okay with the fact that dietary supplements are not required to be tested for safety or efficacy before being sold? Do we mind that manufacturers can conceal ingredients in "proprietary blends," or that dietary supplements are being spiked with pharmaceutical drugs? Would we like to know about potential interactions or long-term side effects, or have a guarantee that the ingredients in supplement bottles are what their labels claim? Should we be concerned that there are no firm requirements for what type of evidence must exist for supplement structure/function claims, or that the FDA is largely prohibited from limiting what combinations or doses of ingredients can be sold? Does it make sense to assume that everything "natural" is harmless, regardless of dose, or that all supplements on the market before 1994 can automatically be assumed to be safe? Can we expect our doctors to be aware of all possible interactions, especially if they themselves might not even know that supplements aren't evaluated by the FDA (let alone if we lie about what we're taking)? Do we trust the supplement industry to regulate itself? Do we really believe this is a matter of personal freedom? Or have we been manipulated into becoming personal advocates for an industry that takes in billions of dollars—from us—a year?

Getting back to the thirteen actual vitamins, far from being outshone by the exotic supplements that they have enabled, their appeal has only continued to grow. Today's supermarket shelves are weighed down by vitamin-enhanced products—the sports drinks, the fortified snacks, the energy bars—that suggest the more a food is fortified with vitamins, the greater its benefits must be. But how much can vitamins themselves improve our well-being? Is more really better? And how about the value of multivitamins, the nutritional insurance policies that many of us take each day? Regardless of how they're regulated, how much can vitamins, whether we eat them in foods or take them as pills, truly boost our health?

9

Foods with Benefits

The vitamin bottles contain reading material with real style to
it, up to and including "para-aminobenzoic acid." I forget just
what marvel "para-aminobenzoic acid" performs; it may be the
Gladness Vitamin, that makes you a bundle of joy, or it may be
the vitamin that wards off premature senility in copper miners.
It doesn't matter. The point is, a title like that makes the
customer feel he is really getting something for his money.
—ROBERT W. YODER, *Hygeia*, APRIL 1942

Despite the estimated 85,000 supplement products for sale in America
and the passion people feel for them, the National Center for Complementary and Alternative Medicine—an NIH-funded research organization that was founded under the assumption that some of these therapies *do*
work has not found benefits for many of the most popular supplements on
the market. Indeed, if you read through the "Herbs at a Glance" section of
NCCAM's website, you'll notice that when the placebo effect is controlled
for, most herbs' possible side effects are nearly uniformly more significant and
better documented than their potential benefits.

For those substances that do have documented effects, one might argue
that perhaps they should be regulated as over-the-counter drugs, since—
given our current regulatory framework—that's the only way to ensure efficacy, safe dosing, consistent potency, and quality. Doing so might reframe the
way we categorize medicine. As Paul Offit, chief of the Division of Infectious
Diseases and director of the Vaccine Education Center at the Children's Hospital of Philadelphia, writes in his critique of alternative medicine, *Do You
Believe in Magic?*, "[T]erms like *conventional* and *alternative* are misleading. If
a clinical trial shows that a therapy works, it's not an alternative. And if it
doesn't work, it's also not an alternative. In a sense, there's no such thing as
alternative medicine."

It's a thought-provoking point—but good luck applying that logic to vitamins themselves. As we've noted, vitamins exist in an odd middle ground between pharmaceuticals and food. As such, vitamins will never be entirely reclassified as drugs—both because doing so has been legally forbidden and because we associate them so strongly with things that are "natural."

And yet our belief in what vitamins can do for us extends so far beyond nutrition that it borders on the *super*natural. Not only do we consider the mere presence of vitamins to be proof of a food's nutritional quality, but we seem to believe that vitamins are safe and inherently beneficial, regardless of dose. Going even further, many people have become convinced that superdoses of vitamins can do things like boost our energy and our moods, even prevent autism and treat cancer. But even people who don't believe in popping hand-fuls of pills can still be enticed by a product category that these beliefs have inspired: so-called nutriceuticals and functional foods (think vitamin-fortified water or sports bars) that claim that their extra vitamins will produce some sort of health benefit above and beyond what could actually be achieved by your diet.

We buy into this aspect of vitamins' mystique every time we treat a cold with a vitamin-enhanced throat lozenge, assume that the addition of vitamins makes a product healthy, or take a daily multivitamin "just in case." Indeed, our beliefs that vitamins do good things for us are so deeply ingrained that we rarely stop to consider where these beliefs come from, or how justified they may or may not be. Does the presence of vitamins really ensure that a food is healthy or nutritionally complete? And what do nutritional scientists actually know about how much of an additional health benefit—beyond preventing nutritional deficiencies—vitamins can provide?

Let's take a step back for a moment to acknowledge just how hard it is to do research about any aspect of human nutrition. With the exception of outright poisons, most foods' effects on health are subtle and take a long time to develop. Eating a tablespoon of trans fat once isn't going to do much; it's only repeating the habit over time that will increase your risk of heart disease. On the flip side, if you have heart disease now, it would be nearly impossible to conclusively prove that it was trans fat that caused the problem. There are just too many variables in play.

Second, the most convincing scientific studies are those that have a control group and an intervention group—that is, one group that is taking (or consuming) the substance whose effects you are investigating, and a similar group that is not. This is the only way to prove that the substance you're studying actually *caused* the effect you observed. Ideally, you'd also be able to give your control group a placebo—an inactive pill that would prevent your subjects from being able to tell whether they were in the treatment group or the control.

With drugs, this is relatively easy to achieve. Most chemicals being studied as drugs do not naturally occur in foods, so you can be confident that your control group is not getting the substance you're studying from their normal diets. It's also relatively easy to make a convincing inactive pill. But since everyone eats food—and since most food-related health effects take so long to develop—it's very difficult to truly determine the effects of any particular diet or nutrient, including vitamins. You couldn't, for example, forbid your participants from consuming any external sources of vitamin C for fifteen years. And if your intervention is a food, it's also nearly impossible to create a convincing placebo. People tend to know whether or not they're eating broccoli.

As if that's not enough, it's tough to get accurate data on what people are actually eating. Not only do we consciously lie (who's really going to admit to eating the entire box of cookies?), but even if we're trying to be totally honest, remembering exactly what we eat—let alone keeping track of how much—is extremely difficult, especially for foods we don't prepare ourselves. As a result, many of our most common assumptions about food and health come from data sets whose foundations, if we dig into the details, look more like sand than cement.

Consider the Nurses' Health Study, a 238,000-person observational trial—meaning that it didn't include any specific dietary intervention or control group; it just tracked participants' habits over time—that is one of the largest and longest running investigations of women's health. Originally designed to investigate the long-term effects of oral contraceptives, the study also concluded that increased folate consumption is associated with a decreased risk of colon cancer, and that a higher intake of green leafy vegetables is associated with a lower risk of cognitive impairment.

Conclusions like this are often used to guide dietary recommendations—not to mention marketing campaigns. But an "association" can't prove a cause-and-effect relationship; it just means that people who reported eating lots of leafy greens also reported fewer cognitive problems, and that the researchers think there may be a link.

Adding to the confusion, these conclusions were drawn from self-reported data in dietary habit surveys that were done, starting in 1980, about once every four years as part of the overall questionnaire sent out to nurses. When I looked up samples online, I found that the questions asked in these surveys—which are an example of what are known as Food Frequency Questionnaires—were amazingly specific. Nurses were asked to indicate "how often on average you have used the amount specified during the past year" on a list of foods that included teaspoons' worth of nondairy coffee whitener; half-cup servings of sherbet or ice milk (ice cream was a separate category); pats of butter "added to food or bread; exclude use in cooking"; shrimp, lobster, or scallops as a main dish; and four-inch sticks of celery. Respondents were supposed to rate their intakes of these foods on a scale ranging from "never, or less than once a month" to six-plus times per day.

"Please try to average your seasonal use of foods over the entire year" say the instructions, which bear a striking resemblance to SAT math problems. "For example, if a food such as cantaloupe is eaten 4 times a week during the approximate 3 months that it is in season, then the average use would be once per week."

As is perhaps quite obvious, Food Frequency Questionnaires can never be taken as an accurate record of an individual person's food consumption. And that's not their purpose. Instead, they're meant to rank people on a continuum (do you eat lots of high-folate foods, or a little?), and their results are meant to suggest associations—note the doubly noncommittal word choice—between various points on the spectrum and certain health conditions. But of course, if you read the headlines in most magazines and newspapers, it's clear that by the time these studies' results make it to the public, those nuances have been lost.

Aware of the many difficulties in performing high-quality nutritional research, I turned to the US military, an organization whose researchers have some advantages compared to their civilian-oriented peers. As James McClung, PhD, nutritional biochemist at the military nutrition division of the US Army Research Institute of Environmental Medicine, explained to me, the military's population of potential volunteers includes basic training recruits, tens of thousands of men and women who have already chosen to exclude themselves from the so-called free-living population. As part of their training, these recruits have committed themselves to spending eight to ten weeks in an extremely controlled environment, following similar schedules with similar physical activities and eating similar foods. As a result, the military's nutritional researchers are able to control for external variables in a way that nonmilitary researchers—who have little control over what their subjects are eating or doing—cannot.

Also, throughout history, militaries around the world have had a strong motivation to do nutritional research: they want to create the best-nourished, strongest fighters possible. The American military's interest in using nutrition to promote health dates back at least to 1861, well before vitamins were recognized or nutritional science was an established field. That's the year when an army surgeon named John Ordronaux published the first known dietary guidelines for American soldiers, with recommendations that were echoed, more than a hundred years later, in the US Department of Health and Human Services' first *Dietary Guidelines for Americans*. His prescient observations included that fresh fruits were preferable to dry or preserved ones, that the woody fibre of vegetables provided bulk as well as nourishment, and that the best soldiers in the world were fed on dark-coloured bread.

We've already noted the military's World War II-motivated push for widespread vitamin enrichment and fortification of food; the American military's current areas of investigation range from the role of nutrients in wound healing and recovery from infectious disease to new flavour- and texture-saving preservation methods that could have applications in the grocery store as well as on the battlefield.

The military's interest in food preservation also isn't new. The process of canning fruits and vegetables in jars was developed in the early 1800s by a French chef named Nicolas Appert in response to Napoleon's challenge to develop a method of creating safe, transportable food for armies on the move. This breakthrough (sterilizing food by heating it in sealed containers) helped enable the first mass-produced processed foods. By World War II, further advances in food preservation techniques and packaging materials—plus the development of synthetic vitamins—liberated grocers from having to stock foods that were fresh. For civilians, these advancements led to the creation of many of the packaged, processed—and often fortified and enriched—products that fill the middle aisles of modern grocery stores and survive indefinitely in our pantries. For the military, they enabled the creation of what today is the primary ration for troops in the field, and my particular area of interest: the Meal, Ready to Eat, more commonly known as the MRE.

Launched in the early 1980s, MREs have been approved for use as troops' only food source for up to twenty-one days in combat or other field conditions where there isn't access to normal cooking facilities. They come in tan plastic pouches that manage to cram a 1,300-calorie meal into a 1.5-pound package. Open a modern-day MRE and in addition to one of a rotating selection of twenty-four entrées, you'll find accompaniments like drink mix, pound cake, snack bread, and chocolate peanut butter. Most MRE pouches also contain an ingenious technological innovation called a flameless ration heater, which lets you heat up your entrée without any need for fire or a microwave. You pour water into the ration heater bag, tuck it and your MRE entrée into a cardboard box, and prop the whole thing against a "rock or something"—that's the military's official terminology from the instructions printed on the package—and several minutes later, your meal is hot.

I was particularly interested in MREs because they reminded me of the "purified diets" developed by late nineteenth- and early twentieth-century nutritional chemists—the man-made, supposedly complete formulas that caused deficiencies in lab animals and, in so doing, prompted the discovery of vitamins. Like those early artificial diets, MREs are engineered creations, designed by some of the most knowledgeable nutritional experts in the country. Everything we know about humans' short-term nutritional needs is incorporated into an MRE—and if there's any compound that has been shown to provide a performance enhancement beyond basic nutrition, the military has every incentive to include it as well. I hoped that MREs might reveal what we know about vitamins' ability to enhance health and performance, as well as how close we are to being able to reverse engineer nutritionally complete food.

And so I made a visit to the ground zero of MRE development: the Combat Feeding Directorate, which is part of the US Army Natick Soldier Research, Development and Engineering Center in Natick, Massachusetts, often referred to as the Natick Army Labs. It's also home to the labs responsible for the development of fire-resistant fabric, new camouflage patterns, and as one person summarized it to me, "pretty much everything that doesn't explode or get shot out of a gun." But the goal of the Combat Feeding Directorate is entirely food-focused: "To ensure that United States Warfighters"—that's the current catchall term for military service members—"are the best fed in the world."

My day began with a tour of the building, a squat structure that was home to, among other things, the ration test kitchen and the Polymer Film Center of Excellence (it works on packaging). Portraits of Louis Pasteur and Typhoid Mary stared down from hallway walls, and informational posters addressed subjects like "Evaluation of High Pressure Processing of Wet-Pack Fruits" and "Measuring the Effects of Nano and Synergistic Formulations on Curcumin Availability/Absorption Using Saliva and Buccal Permeability Experiments."

Eventually my guides and I arrived at the Warfighter Café, a meeting room lined with glass display cases containing a historical assortment of American rations. My Combat Rations Team hosts, team leader Robert Trottier and dietitian Julie Smith, explained that the first iteration of MREs—which became standard issue in 1986—were not well received. Designed without troop input (and without the heaters), they had a limited and monotonous selection of entrées, and earned nicknames like Meals Rejected by Everyone and Meals Refusing to Exit. One story told is that after the first Gulf War, Colin Powell pointed at an MRE and said, "Fix it." A Department of Defense booklet puts it a bit more diplomatically: "Feedback from Operation Desert Shield/Storm suggested that Warfighters would consume more if their preferences were taken into consideration." Either way, the result was what I was there to experience: palatable and varied rations, designed with constant user feedback, that are "Warfighter Recommended, Warfighter Tested, Warfighter Approved."

Surveys have revealed certain guaranteed crowd-pleasers—Warfighters, for example, like things spicy. (I do not. When Smith saw me choking on a Jalapeño Pepper Jack Beef Patty, she offered me a palate-cleansing piece of Wheat Snack Bread, kindly waiting for me to swallow before revealing that it was two and a half years old.) Pot roast is consistently a favourite. Tabasco sauce is a must. And as Smith and Trottier both affirmed, "the Warfighter loves the cheese spread."

But beyond that, Smith and Trottier explained, Warfighters' tastes can be fickle.

"Buffalo chicken used to be high on the ratings, but now it's down and chipotle is way up," Smith told me.

"Which MRE entrée has been the least popular?" I said, nibbling my toddler-aged snack bread.

"You mean in terms of complete and utter distaste from every soldier that ever ate it?" asked Trottier.

Smith answered so quickly that she nearly cut him off. "The omelet," she said as Trottier nodded emphatically. "That was the biggest failure in the last ten years."

But good taste is only one of the requirements that MREs must fulfill. Since a Warfighter might have to carry multiple MREs in his or her pack, they are designed to be as energy-dense and lightweight as possible—an approach similar to that advocated by Wilbur Atwater, the nutritional chemist who, at the turn of the twentieth century, recommended that the poor maximize their dollars by prioritizing high-calorie, high-protein foods. MREs also have to stay shelf-stable (that is, not spoil) for six months at 100 degrees Fahrenheit and three years at 80 degrees, can't ever require any refrigeration, and—as if that's not enough—must be able to withstand being airdropped.

These requirements knock out many of Warfighters' most requested foods. You can't have macaroni and cheese, because the heat, pressure, and long cooking time of the sterilization process—which is called retorting and basically is the same as canning—would turn it into a gloppy mess. You can't have normal sandwiches, because sandwich bread is too moist: water promotes bacterial and mold growth, so most of the MREs' bread and cake products are extremely dry and crumbly, their texture similar to that of a Pop-Tart crust. Also, the fillings would make them soggy—who would want a three-year-old peanut butter and jelly sandwich? It's difficult to preserve pizza in a pouch. Even chocolate is a challenge because it melts. The solution is "pancoated chocolate discs"—what we would call M&M's. Trottier described MREs as the military's "bread and butter," but in reality, they can't contain either.

The true MRE pariahs, though, are fruits and vegetables. They fail on all counts: they're high in moisture, which makes them heavy and prone to spoil, they contain few calories, and they get mushy when retorted. While produce makes cameo appearances, say in Beef Roast with Vegetables or Southwest Beef with Black Beans, it never has a starring role. Instead, MREs' base ingredients rely heavily on sugars and refined flours.

As a result, an unfortified MRE would not be a very nutritious meal, at least where vitamins are concerned. The only vitamin in white sugar is riboflavin, which is present at a (non-) whopping 0.038 milligrams per cup. And refined wheat flour, which is derived from the centre of the wheat grain, isn't much better. Since its sole purpose is to provide energy for the embryonic wheat to germinate (like the white part of rice, it's a little gas tank), it's basically pure starch; what little fibre it contains is nutritionally poor.

What's more, whatever vitamins do exist in the MREs' raw ingredients are unlikely to survive the level of processing necessary to render them shelf-stable for the lengths of time required. Vitamins, like emotionally sensitive people, each have their own vulnerabilities. They can be destroyed by heat, air, moisture, pH, light, or even simply the passage of time. Vitamin C is particularly challenging: it's sensitive to *everything*. This is why many fortified foods and vitamins contain "overages" (that is, more micronutrients than are on the label)—to comply with regulations, companies need to make sure that by the time their products reach your table, they still have at least as many nutrients as their labels claim.[1]

The destruction of vitamins caused by processing is a problem because, in addition to tasting good, weighing little, and resisting bacterial invasion, there's another requirement that all MREs must fulfill—and this is where the analogy with nineteenth-century artificial diets really comes into play: they have to satisfy service members' every nutritional need. This means that MREs must meet the Military Dietary Reference Intakes (MDRIs) for all macro- and micronutrients. Set by the surgeon general of the army, the MDRIs are based on the civilian versions of the Dietary Reference Intakes from the Food and Nutrition Board, which contain the most updated set of RDAs—but they take into account lifestyle factors common in the military, like intense physical exercise and being seventeen to fifty years old.

---

[1] Stable vitamins typically require overages of about 15 percent, but unstable vitamins like thiamin, A, and C often require overages of more than 50 percent. This means that even if nutritionists were able to determine the "perfect" amount of each vitamin for each person, it would be extremely difficult—if not impossible—to guarantee that that precise amount would end up in your food.

The military isn't allowed to issue dietary supplements to its members in pill form (though many take them on their own—and many bases have supplement stores like GNC). So like many food companies, the military fulfills these requirements by fortifying and enriching MREs with minerals and synthetic vitamins. A chart on a box of garlic mashed potatoes in one of my MREs explained that its accompanying crackers had been fortified with B1, B2, niacin, B6, and calcium; the beloved cheese spread was fortified with vitamins A, C, B1, and B6. The chart, which was titled "Nutrition: A Force Multiplier," put Xs next to parts of the MRE that "should always be eaten"—a nod to one of ration designers' greatest fears: that Warfighters will ignore their carefully crafted menus and exist solely on pound cake.

The military also isn't allowed to issue performance-enhancing drugs, whether as pills or in food. Ration designers cannot, for example, spike the mashed potatoes with steroids. However, the military is greatly interested in dietary compounds found naturally in food that may optimize human health and cognitive and physical performance, including anti-inflammatory substances such as omega-3 fatty acids and curcumin (a substance in turmeric) and forms of long-acting—and therefore energy-sustaining—carbohydrates. Zapplesauce, for example, is applesauce that's been fortified with maltodextrin, a complex carbohydrate used in sports gels like GU. And this focus on functional ingredients brings me to a very important point. If you ever visit the Combat Feeding Directorate yourself, and you are treated to a ration-themed lunch buffet, please, I beg you: Do not eat the caffeinated meat sticks.

It might seem like a good idea at the time. You'll be sitting in the Warfighter Café with an array of white plates arranged before you on a camouflage-patterned tablecloth, each holding bite-size samples of rations for you to try. There will be Meatballs with Malfada Pasta and Orange Cake spiked with omega-3 fatty acids, and you will have just been shown a clip from the TV show *Rock Center*, in which Brian Williams praises the caffeinated meat sticks as "an invention that will be ranked right up there with the iPad and the can opener" before popping several into his mouth. You will have heard his host nervously tell him that each small stick contains the caffeine equivalent of about a cup of coffee ("so you might want to go slow") and seen Williams shrug off her warning with a characteristically adorable grin. And you will think, I can keep up with Brian Williams. But trust me, you don't want to.

I'd got up at four thirty that morning, and the day had continued non-stop. I had already visited the ration test kitchen, where I sampled new menu items under development, including fruit cocktail and a maple sausage wrap, and rated them on a hedonic scale that the Combat Feeding Directorate uses to evaluate potential new menu items (rations won't make it into rotation unless they score at least a 6 out of 9). I had learned about the challenges of creating flexible packages of Tabasco sauce—so beloved by Warfighters that it simply must be included, but so corrosive that until recently it had to be distributed in miniature glass bottles. I'd even tried "tube food": special rations that U2 pilots suck down via a tube when they're flying on extended reconnaissance missions at altitudes so high that they can't take off their pressurized suits and helmets (they're known as "foods with altitude"). The beef Stroganoff was not the finest that I've had, but the tubed apple pie was genuinely delicious.

It was fascinating, but I was exhausted. So when someone placed a bowl of meat sticks in front of my seat at the lunch table, I began gobbling them up like peanuts. Since these particular samples had been cut into snackable, bite-size chunks, I don't know how many full sticks I consumed. Six? Seven? Whatever. Within minutes, I understood why the commercial version is called Perky Jerky.

I've never experienced a caffeine buzz like this. I didn't just feel awake. I felt inspired, confident, and borderline euphoric; the afternoon's first post-lunch session, about microwave sterilization and osmotic drying, left me feeling as pumped as a workout. And the jerky's effects were long-lasting too. When I called my husband that evening, more than eight hours after my last bite, I unleashed a torrent of speech so fast and so hyper that he interrupted me to ask if something was wrong.

When I tried to get to sleep, it became painfully clear why the military issues packets of caffeinated gum in its First Strike Ration packs, which are designed for the first seventy-two hours of combat.[2] (Its commercial tagline is

---

[2] Interestingly, while the military can't add supplements in pill form to its rations, it *can* add things like fortified gum, chews, powdered drinks, and gels—another example of the blurring of the lines between supplements, foods, and drugs.

"Stay Alert, Stay Alive.") At the same time, however, caffeine's inclusion in rations—it's also incorporated into energy bars and chocolate pudding— also struck me as kind of weird. Caffeine isn't a nutrient; it's a stimulant. Its existence in rations seems to be a holdover from the days when they also included alcohol and cigarettes. So if caffeine is allowed, what about extra vitamins? Given that rations are already fortified with synthetic vitamins, this would be both totally acceptable and presumably—if popular logic is to be believed—beneficial.

Sure enough, an initial glance at my MRE made it seem like there might be some benefit to higher-than-normal doses of vitamins—a graphic on one of the boxes proclaimed "Fortification Provides you the Additional Edge to Maximize Your Performance," suggesting both that the military had found extra vitamins to be performance-boosting and that those extras must be in my garlic mashed potatoes. But when I looked at the actual nutritional information on the package, I noticed that my potatoes didn't contain an unusual amount of any vitamin. Instead—like all MRE ingredients—they were fortified just enough to help bring me up to the military's recommended nutritional standards, nothing more.

It turns out that this is a deliberate omission, because the military—like civilian researchers—has not found any short-term benefit from vitamin superdoses. In fact, caffeine is the most certain performance-enhancing ration ingredient that the military has found.

Of course, given its population and purpose, the military is more interested in investigating short-term benefits than it is in the prevention of age-related, chronic diseases like cancer that take years to develop. But when I looked into the work that's been done on the effect of high doses of vitamins on long-term disease prevention, I was similarly surprised. Despite our desire to think of vitamins as health superheroes, their actual powers—when it comes to doing things beyond preventing or curing outright deficiency—do not appear to live up to our dreams.

Scientists' inability to establish evidence in support of high-level vitamin supplementation—whether short term or long term—is not for lack of trying. Even before the World War II-era push for thiamin, vitamins were capturing the attention of both researchers and the public as possible panaceas for a panoply of ailments, and individual vitamins have been cycling in and out of fashion ever since. The most famous example is undoubtedly vitamin C, megadoses of which, as noted earlier, the chemist Linus Pauling erroneously claimed could cure everything from the common cold to cancer. But when it comes to the unfulfilled promises of vitamin megadoses, and the possible dangers of believing in them, beta-carotene—a precursor to vitamin A—stands out as a cautionary tale.

As you know from the story of golden rice, our bodies can convert beta-carotene into vitamin A. But scientists think that beta-carotene may also play important health roles of its own. In 1981, renowned British epidemiologist Richard Peto published a review article in *Nature* summarizing numerous observational studies in humans whose findings suggested that people who ate more orange-red vegetables—and therefore consumed more beta-carotene—had a lower incidence of many forms of cancer. This beta-carotene hypothesis tied in intriguingly to the idea, also promulgated by Peto around the same time (and still under investigation), that some 35 percent of preventable cancers could be attributed to unspecified dietary factors.

Many researchers speculated that the beta-carotene was preventing cancer by acting as an antioxidant, meaning that it was protecting cells from reactive molecules called free radicals that lurk in everything from cigarette smoke to air pollution. Judging from the way they're referred to in magazines and television news segments, free radicals are agents of pure evil: left unchecked, they will probably kill you in your sleep.

The actual definition of free radicals sounds decidedly less scary. They're molecules that are missing an electron—the negatively charged component of an atom—from their outer shells. While that might not sound like a big deal from a human perspective, it's the molecular equivalent of waking up one morning and realizing that your arm has gone missing. Chances are, you'd want a replacement. Free radicals are similarly desperate to replace their missing electron. So they commit crimes of opportunity: they steal one from a nearby molecule, which then becomes unbalanced. Then that unbalanced molecule steals an electron from another molecule, which makes *it* unbalanced, and so on and so forth, creating a chain reaction in the body (or other organism) that, if left unchecked, can cause damage and mutations in DNA and other important cell components that can lead to diseases like cancer and heart disease. This is known as oxidative damage, or oxidative stress.

Antioxidants like beta-carotene and vitamins C and E are special sorts of molecules that are able to donate an electron without becoming unstable themselves, thereby stopping potentially dangerous chain reactions before they get out of hand.

That's not to say that free radicals are universally bad, however. They instigate health-preserving reactions as well, and play crucial roles in our immune systems (our bodies instigate oxidative reactions to destroy bacteria and other pathogens, for example). Recent research also suggests that they may have some sort of beneficial effect after exercise. As is so often true, the trick is to find a healthy balance—the problem is that at the moment, no one knows what that healthy balance actually *is*.

In any case, at the time of Peto's paper, the evidence for beta-carotene from animal and in vitro studies was, in the words of one summary of the time period, "overwhelming," providing "the most persuasive evidence available in the diet-cancer epidemiological literature for a protective association, in terms of both magnitude and consistency." Indeed, of the more than 125 observational studies on beta-carotene and cancer risk performed between 1980 and 1990—on populations that included smokers, nonsmokers, different racial groups, men, and women—most suggested a 50–150 percent greater risk of cancer in the people with the lowest levels of dietary beta-carotene.

But, as noted, the problem with observational trials is that they don't have a control group. In this case, that would mean a comparable group of people who'd been randomly selected to *not* take (or consume too much) beta-carotene, which would help determine whether beta-carotene was actually responsible for the observed effect. Instead, most researchers had simply looked at people's self-reported dietary habits and compared them with their rates of cancer—which meant the possibility for confounding variables was high. And so when Peto's paper, provocatively titled "Can Dietary β-Carotene Materially Reduce Human Cancer Rates?" was published, it prompted a wave of randomized, placebo-controlled clinical trials—the gold standard for research—for beta-carotene and other antioxidants like vitamin E, vitamin A, and selenium. It took years for the studies to run their course, but finally, between 1993 and 2000, five large intervention trials published their results.

Data from the first trial, which was performed in Linxian, China, and published in 1993, looked promising. Giving five years' worth of vitamin E, beta-carotene, and selenium supplements (at amounts of about one to two times the then-current RDA) to nearly thirty thousand mostly nonsmoking men and women seemed to result in about a 9 percent lower incidence of death, mostly cancer-related—a protective effect that supported the hypothesis drawn from the observational studies. The researchers noted that their subjects had "persistently low intake of several micronutrients" to begin with, and that this might have accounted for some of the improvement, since, as we saw earlier, vitamin A deficiency increases the risk of death. Nonetheless, the findings were promising enough that the researchers believed they should "stimulate further research to clarify the potential benefits of micronutrient supplementation."

Next to be published was the randomized, double-blind (meaning that neither the researchers nor the participants know who's taking what), placebo-controlled Alpha-Tocopherol and Beta-Carotene (ATBC) Cancer Prevention Trial. It began in 1985 and was designed to investigate the effects of various combinations of high doses of alpha-tocopherol (a form of vitamin E, which is also an antioxidant) and beta-carotene on just under thirty thousand heavy-smoking, middle-aged Finnish men.

Previous observational studies had suggested that beta-carotene might be particularly beneficial for smokers, since cigarette smoke is a major source of free radicals. But that's not what happened. Instead, when the study's results were published in 1994 (after the men had used the supplements for between five and eight years), they showed a 16 percent *increase* in the incidence of lung cancer among the men taking the beta-carotene supplements, and an 8 percent higher incidence of death than the placebo control. This was such a shocking result that some scientists assumed that the trial must have been flawed. But eight years later, the intervention group's risk levels had returned to the baseline—suggesting that the supplements were indeed responsible for the effects.

Next up was the CARET study (the veggie-friendly acronym for the Beta-Carotene and Retinol Efficacy Trial), which was designed to test the effect of high doses of beta-carotene and vitamin A on the incidence of lung cancer, other cancers, or death on about eighteen thousand people with a history of smoking or asbestos exposure. It brought additional bad news: designed to run for about ten years, it was stopped nearly two years early when a check of the data showed a 28 percent *increase* in lung cancer among the study subjects taking the supplements. In this case, six years later a follow-up study on the same population found that female smokers who had received vitamin A and beta-carotene were still about 35–40 percent more likely to develop lung cancer or die of any cause than women who had not received the supplements.

In the words of Serge Hercberg, a researcher who wrote a history of the study of beta-carotene and cancer, "It was as if Dr. Jekyll became Mr. Hyde. Beta-carotene not only was considered not to be protective but was also potentially deleterious, with an increased risk of lung cancer among smokers. . . . Nevertheless, the beta-carotene experience seems to have had minimal impact on the dietary supplement industry."

Finally, the last two of these five large-scale, randomized, and controlled trials were published—the Physicians' Health Study I (PHS I), which treated 22,000 American primarily nonsmoking male physicians with high doses of beta-carotene, and the Women's Health Study, which did the same with nearly 40,000 healthy female health-care professionals. Both found no benefit *or* harm from beta-carotene supplementation, even among the smokers. Overall, the results were inconclusive at best: of the original five large studies on beta-carotene, one had been slightly positive, two had been alarmingly negative, and the other two were neutral. The general consensus today is that large doses of beta-carotene are not beneficial—and are potentially dangerous, especially for smokers.

Why were the results of these beta-carotene studies—all of which were well designed—so contradictory? Was it because the people in the research populations were consuming different levels of beta-carotene in their diets or had different levels of vitamin reserves to begin with? The participants in PHS I and the Women's Health Study, for example, were likely far better nourished than the Linxian population. Could beta-carotene be beneficial for nonsmokers and poorly nourished people, but neutral or dangerous for smokers or people who may have already begun to develop cancer? Could it switch from functioning as an antioxidant (stopping chain reactions) to a *pro*-oxidant (instigating them), depending on dose and what other chemicals are present? Is it possible to consume too *many* antioxidants? If so, where is the line? Are there differences between how our bodies react to beta-carotene in supplements and beta-carotene as it naturally exists in food? Were there other differences between the research populations that could have affected the results? No one knows for sure.

The disappointing results of the beta-carotene trials were difficult for many people to accept, in part because they contradict our belief that if a small dose of a vitamin is good, a bigger one must be even better. It's a tempting theory, and one that—thanks to supplements—is very easy to act on. But while our more-is-better philosophy has a satisfying intuitive logic, it doesn't hold up to common sense. Even water can kill you if you drink too much of it.

Instead, the results of nearly all randomized, controlled studies of vitamin megadoses—our would-be nutritional magic bullets—have suggested the opposite of what we want to hear: that the healthiest and safest doses of vitamins are the ones naturally found in food. Just as Richard Peto's observations launched a wave of research on beta-carotene, these studies suggest it might be time to shift our focus away from superdoses and more toward moderation—a Goldilocks approach that eschews excess in search of a middle ground.

Two well-designed studies in particular support this idea. In 2004, the Supplémentation en Vitamines et Minéraux Antioxydants (SU.VI.MAX) study was published. For seven and a half years, thirteen thousand subjects had taken supplements that contained much lower doses of beta-carotene (and several other antioxidants) than were used in the previously mentioned studies. The study concluded that "an adequate and well balanced supplementation of antioxidant nutrients," given at doses that would be achievable with a diet rich in fruits and vegetables, "had protective effects against cancer in men."

Then in late 2012, the multivitamin/cancer results of the Physicians' Health Study II (the second phase of the aforementioned Physicians' Health Study) were published in the *Journal of the American Medical Association*. The study, which was randomized, double-blind, and controlled, and followed fifteen thousand middle-aged male physicians, did not find any improvement in cancer risk from *high* doses of vitamins C, E, or beta-carotene. Nor did it find any benefit from daily multivitamin use on the risk of major cardiovascular events or cognitive decline. But the men in the multivitamin branch of the study, who took a daily Centrum Silver for an average of eleven years, did have a modest but statistically significant reduction in total cancers—a reduction, said the study's principal investigator, John Michael Gaziano, MD, of about 10–12 percent.

The results of PHS II received far more attention from the American press—and public—than those of the SU.VI.MAX trial. Maybe that's because the Physicians' Health Study's results were published more recently. Maybe it's because SU.VI.MAX was French. Or maybe it's because, by looking specifically at *multivitamin* use (as opposed to SU.VI.MAX's attempt to use synthetic vitamins as a way to replicate doses found in food), the Physicians' Health Study II provided a sense of scientific justification for a product many of us have used since childhood.

Today, more than half of Americans and around a third of Britons report taking dietary supplements, many of which are multivitamin/multimineral products. (I will succumb to the common habit of just calling these "multivitamins.") To their fans, these daily multivitamins are an insurance policy, a habit as essential to responsible adulthood or parenthood as maintaining collision coverage on your car. Indeed, this idea has become so ingrained in our culture that, much like occasional churchgoers, even those of us who aren't true believers *still* take a multivitamin "just in case."

Our embrace of multivitamins reflects another nutritional philosophy that was pushed by the military and the government during World War II: that it's possible to be well fed calorically but nutritionally deficient nonetheless. (Today, this phenomenon is referred to as "hidden hunger," and is estimated to affect two billion people in the developing world.) If there's no way to identify which of our nutritional needs are not being met—or if we *know* that we're feeding ourselves and our children junk—then we'd better cover our bases by packing in as many extras as we can.

Given our devotion to our multivitamins, you may be surprised to learn that the US Preventive Services Task Force, the American Cancer Association, the American Heart Association, the American Diabetes Association, and the American Academy of Family Physicians, among other respected health organizations, do not recommend that healthy people with no nutritional deficiencies take multivitamin supplements. That's not because multivitamins are harmful in and of themselves, which seems unlikely for most people. It's more that, while multivitamins can help prevent deficiencies and some vitamin supplements have earned broad scientific support (folic acid for women of childbearing age to prevent neural tube birth defects, calcium and vitamin D to prevent bone-thinning in postmenopausal women, and vitamin B12 for vegans and people over fifty), there is very little—if any—evidence that multivitamins confer additional benefits that cannot be obtained through a healthy diet. Instead, if we rely solely on multivitamins to fulfill our nutritional needs, we'll be missing out on all the other important compounds in natural foods that those pills (and fortified products) don't contain.

Also, given how many of our foods are fortified and enriched, the debate over whether to take a multivitamin is somewhat redundant: many of us are *already* taking a multivitamin, in the form of fortified and enriched food. Indeed, people who eat a lot of fortified and enriched products are likely consuming more vitamins than people who get their vitamins from foods that naturally contain them.

All told, this is why America's leading health organizations recommend eschewing multivitamins in favour of consuming a nutrient-dense diet, and why scientists and nutritionists repeat, again and again, that if the healthiest doses of vitamins and other micronutrients appear to be those found in food—and if food contains other chemicals that are likely beneficial to our health—then we should stop taking pills and *just eat food*. "The message is simple," stated a strongly worded 2013 editorial in the *Annals of Internal Medicine*. "Most supplements do not prevent chronic disease or death, their use is not justified, and they should be avoided. . . . Enough is enough."

And yet despite these pronouncements and the flurry of news and magazine articles that they typically inspire, we do not take their underlying messages to heart. We not only continue to take our multivitamins (and in some cases, superdoses) but also eat diets that are remarkably similar, in terms of their dependence on synthetic vitamins, to MREs. What's more, whereas the military has only approved MREs as a sole food source for twenty-one days, we use these products as cornerstones of our daily diets, lulled into believing that their synthetic vitamins make them a worthwhile substitute for naturally nutrient-dense foods. Troops rely on this human-designed food in the field because they don't have access to anything else. But given that we're free to choose whatever we want at the grocery store, why do *we*?

From the discovery of vitamins to the development of MREs, the past century has seen huge advances in our quest to understand the intricacies of human nutritional needs. Today, we know how to prevent vitamin deficiency diseases, and have fortified and enriched our food supply to protect ourselves from them. We know how to produce bulk synthetic vitamins by the ton. Like our historical predecessors, our hope is that if we continue breaking down food into smaller and smaller parts, we'll eventually crack the code of human nutrition. We're driven to keep working on this problem not just because of our human desire for understanding, but also because we hope that doing so will allow us to find an easy, ideally effortless way to achieve perfect health.

But solving the mysteries of human nutrition is a bit like peeling an onion: the more layers you pull back, the more layers you find underneath. As the limitations of MREs and the continued uncertainties over the RDAs themselves demonstrate, even a hundred years after the failure of purified diets and the discovery of vitamins, we still don't know how to reverse engineer perfect food. Nature is simply too complex. Indeed, it might well be an impossible goal—for how could we ever be confident that we weren't missing something?

This question applies not just to MREs but to infant formula, fortified breakfast cereals, meal replacement shakes, multivitamins, and every other human-designed food we come into contact with each day. Indeed, the very idea that we might know precisely which chemicals each of our bodies needs (and in which quantities and combinations) brings to mind the innocence— and arrogance—of chemists at the turn of the twentieth century whose artificial foods failed because they didn't know about vitamins. Like our predecessors, we, too, are likely leaving out compounds whose importance we don't yet recognize or understand. The story of vitamins is definitely one of scientific triumph, but it's also a cautionary tale, a reminder that the most important issue in nutrition isn't just what we know; it's what we *don't*.

10

The Nutritional Frontier

[O]nly the surface has been scratched
in relating nutrition to health.
—WALDEMAR KAEMPFFERT,
"WHAT WE KNOW ABOUT VITAMINS,"
New York Times Magazine, 1942

One afternoon, toward the end of my research, I found myself standing in a field in California, nibbling on an alfalfa leaf. The scene might have looked pastoral, but its purpose was scientific: I was trying to gain firsthand experience with an area of research that, like vitamins at the turn of the twentieth century, represents the cutting edge of nutritional science today.

The alfalfa was in a production field at Nutrilite, a plant-based supplement company that has made a business—one that recently brought in close to $4.7 billion in annual sales—out of extracting chemical compounds from plants and packaging them into dietary supplements. My goal was to extract what they've extracted—not because I believed we should all be eating alfalfa in pill form, but because I was intrigued by Nutrilite's philosophy toward nutrition (and indeed, its business model). It's an approach that relies on respecting the mysteries of plants.

Like all living things, alfalfa is packed with chemical compounds that can affect human health. In the case of plants, these substances are called bioactive phytochemicals (*phyto* comes from the Greek word for "plant"). Most vitamins are phytochemicals; as we've discussed, many vitamins do similar things for plants that they do for our own bodies, and the only human vitamins that plants cannot produce in notable quantities are vitamins D, A, and B12. (It's also worth noting that animal products contain all vitamins in reasonable quantities except for vitamin C.)

It's the potential actions of other groups of phytochemicals, however, that are responsible for all the headlines about "superfoods" that pop up regularly in newspapers and magazines alongside colourful photographs of blueberries and pomegranates. Indeed, these chemicals' names alone make them perfect candidates for nutritional hype, difficult to pronounce and scientific-sounding enough to appear authoritative. Just like Elmer McCollum's early twentieth-century vitamin roundups in *McCall's* magazine, the simplicity of the resulting nutritional charts and infographics imply that the science of how phytochemicals affect our bodies is cut and dried. But don't be fooled. There are thousands of phytochemicals in the foods that we eat, relatively few of which have been rigorously studied in humans, and none of which we fully understand. Consider this list of just *some* of the phytochemicals under study: flavonoids, flavonols, flavanones, isoflavones, anthocyanins, anthocyanidins, proanthocyanidins, tannins, isothiocyanates, carotenoids, allyl sulphides, polyphenols, and phenolic acids. As journalist Michael Pollan puts it, we still have not gazed into the soul of a carrot.

None of these phytochemicals is likely to become known as a new vitamin, both because there are no obvious nutritional deficiency diseases associated with any of them and because of the historical haphazardness of the term "vitamin" to begin with. Nonetheless, the idea that plants (and animal) products might contain important chemicals beyond their vitamins and minerals is a powerful one.

The founder of Nutrilite, which is one of America's oldest dietary supplement companies, was one of the first people to capitalize on the idea. He was an itinerant entrepreneur named Carl Rehnborg, who'd been put in charge of launching Carnation Evaporated Milk in China (an unlucky assignment, given that most Chinese people are lactose intolerant). Living near Shanghai's French Concession during a period of political unrest in the 1920s, Rehnborg noticed that the health of many of his fellow expats, who were also living in isolated enclaves, appeared to be deteriorating. Already interested in the connection between nutrition and health, Rehnborg sensed a business opportunity—what if he could create a concoction that would revitalize his neighbours? So he took what he considered to be the natural first step in product development: he began experimenting on his family.

"I bought yeast in one-pound cakes and used rice polishings and things of that sort, which were added to the diet," he wrote. He used an old grinder to pulverize bones and added them to soup. He supplemented recipes with potato skins, which he correctly guessed contained vitamins, and dusted entrées with powdered shells for calcium and phosphorus. According to his own recollections, some of the concoctions were "the most God-awful messes, concentrates of milk, kelp, fish oil, wheat germ oil, liver, alfalfa, watercress, yeast, and parsley." He even developed a beverage that derived iron from an infusion of rusty nails.

While Rehnborg's Carnation milk business in China failed, he continued to be interested in nutrition. So when the political situation for foreigners became threatening, he returned home to the United States and continued his experimentation. In September 1934, he introduced one of America's first multivitamin and multimineral food supplements. It was based on a liquid concentrate of plant extracts—namely watercress, parsley, and alfalfa, which he considered a wonder food.

Though Rehnborg was fast to incorporate synthetic vitamins into his products as they became available, his company's emphasis was always on plants. You can see this philosophy today in one of Nutrilite's top sellers, its Double X Vitamin/Mineral/Phytonutrient product. In addition to providing "a powerful blend of 12 vitamins and 10 minerals, most far exceeding the Daily Value," Double X also includes a blend of concentrates from twenty plants, including cranberries, apples, pomegranates, and kale. At about $85 for a thirty-one-day supply of tablets, Double X caters to our desire to get the benefits of fruits and vegetables without actually having to eat them.

While our understanding of phytochemicals other than vitamins is relatively primitive, we do have hypotheses for some of their purposes in plants. The bright orange-red of beta-carotene is helpful in absorbing light energy from the sun and attracting pollinators (think of a marigold); it also acts as an antioxidant. Anthocyanins, which are chemicals that give many berries their dark blue, red, or purple colours, are thought to act as sunscreen for the plants by absorbing damaging wavelengths and to entice animals to eat the fruit (thereby spreading the plants' seeds). Other compounds, like flavonols, appear to protect plants from pathogens like insects or disease—and these defensive roles account for the astringent or bitter taste of many phytochemicals. Some plants   can   even   release   toxic   compounds   to   sabotage   their neighbours—botanical chemical warfare! And, of course, as the deadly effects of hemlock or poisonous mushrooms make clear, not all phytochemicals are good for us.

To *fully* understand what phytochemicals do for plants (or humans, for that matter), we first need to identify what they actually are. A study published in 2000 in *Nature* provided a great example of how enormous a task this is. It found that 100 grams of apple produced the same amount of antioxidant activity as 1,500 milligrams of vitamin C. (To put this in context, the current RDA for most adults is 90 milligrams of C per day, and a medium apple weighs about 200 grams.) But 100 grams of apples only contains 5.7 milligrams of vitamin C—which means that most of the apples' antioxidant activity must not be produced by vitamin C at all; instead, it's likely due to some combination of phenolic acids and flavonoids, and/or other phytochemicals that we don't yet know about.

The puzzle of identifying these phytochemicals is what Kevin Gellenbeck, a senior research scientist in concentrate development at Nutrilite, refers to as the "dirty chromatogram."

"Not dirty in that way," he said when he noticed my expression. He was telling me about the printouts he gets when he passes a sample of a fruit or vegetable through a chromatograph, a machine that separates different chemical compounds in a sample and charts them as peaks on a graph.

"If I were to do an assay of pure, synthetic vitamin C, it'd be beautiful," Gellenbeck explained. "You'll get this one clean peak—there's nothing else there. But if I do the same thing with acerola cherry extract [a type of cherry that's particularly high in vitamin C], you'll see a peak for vitamin C, but you'll also see all this other stuff. Those are all the other phytochemicals in the plant that came out of the same extraction."

"So how do you know what those other things all are and what they're doing?" I asked him.

His answer was refreshingly simple: "We don't."

As a direct-sales company whose products are never offered in stores, Nutrilite places an emphasis on cultivating relationships with its salespeople. So it's created something called the Nutrilite Brand Experience: several days' worth of customized health evaluations and educational sessions that are designed to give Nutrilite salespeople from all over the country firsthand experience with the products they're selling. They offered me a chance to participate myself, so my morning had already included blood draws and timed sit-ups tests, the results of which—in addition to being evaluated and explained to me by a medical doctor—would be used to provide personalized product recommendations. My tour guide for the day, Takeshi Saito, was the official ambassador and educator for the Nutrilite Brand Experience. Easy to laugh and constantly smiling, he wore a white lab coat adorned with an embroidered alfalfa leaf.

It was an appropriate touch of flair: Saito is the physical embodiment of everything the brand is supposed to represent, a vision of exuberance and health. He is also a genuine believer in the company's products and philosophy. When I later asked him what supplements he took each day—a question I was trying to ask everyone I interviewed—his already cheerful eyes lit up. "Oh!" he said. "I take a lot. There are so many things I don't even want to start."

As its revenues indicate, Nutrilite has moved far beyond the days when Carl Rehnborg hand-processed alfalfa plants gathered from nearby farmers and packaged and mailed the pills himself. But it still aims to create supplements that combine the "best of nature with the best of science," a phrase I heard repeatedly on my tour. That means figuring out how to remove the water, fibre, and sugar from fruits and vegetables and concentrate the remaining substances—the vitamins, the minerals, and the other phytochemicals—into a tablet, and then boosting them with extra synthetic micronutrients if Nutrilite finds nutritional research that it thinks demonstrates a benefit. Nutrilite's Vitamin C Plus product, for example, contains synthetic vitamin C in addition to concentrates from lemons, oranges, grapefruits, and acerola cherries. Rather than try to reconstruct a carrot's soul, Nutrilite's approach is to concentrate it, put it into a tablet, and add some extra stuff for good measure.

I knew that Saito couldn't give me conclusive answers about what, precisely, the plant extracts in Nutrilite's pills were doing in my body, because no one actually knows. Nor is it guaranteed that plant extracts concentrated in pills will act the same as when they're found in food, or even that they'll survive the pill-making process in active form. But I was still interested in finding some tangible way to experience the inner mysteries of vegetables and fruits via Nutrilite's collection of powdered extracts. So I decided to ask if I could do the same thing I'd done with the vitamins at DSM.

When I broached the idea of sampling the powders, Saito was a bit confused.

"Lick?" he said, furrowing his brow. "You want to *lick* the raw ingredients?"

Technically I'd used the word "taste," but sure, I told him, licking would be fine.

"You can look at them and touch them," said Saito. "But if you want to lick your own finger, you're at your own risk."

When I asked him again later in the tour, he reiterated his concerns.

"The powders are for the display," he said. "You're welcome to touch them, but I don't know about licking because a lot of people have touched them, too." It was clear I had put him in an awkward position—that of friendly brand ambassador aiming to please versus that of Nutrilite representative worried about potential litigation from a food-poisoned journalist.

Saito tried his best, but once I saw the display of powders, there was no stopping me. An apothecary cabinet sat on the table, its shelves lined with bottles of colourful powders that were each labelled with an illustration of the corresponding fruit or vegetable. Alfalfa was there, joined by asparagus, broccoli, marigold, parsley, and sage. There was guava (which was yellow) and green tea (which, oddly, was pink). One was simply labelled "Bioflavonoids."

I started with the blueberry, a clumpy purple powder that smelled and tasted only vaguely sweet. Tomato was a relatively flavourless deep crimson, slightly gritty on the tongue and so saturated in colour that I could imagine using it as a dye. Spinach was a slightly bitter green powder. Marigold left my palm orange and tasted a little like iodine. Alfalfa reminded me of hay. Bioflavonoids turned out to be a citrus-scented combination of dehydrated orange and lemon peel. The variety of flavours and textures seemed an indication of just how many phytochemicals there are.

"You're the first guest who's given me feedback," said Saito. He appeared horrified but also intrigued. And indeed, by the time I got to the apple, I'd convinced him to try a sprinkle.

The following is a direct transcript:

Me: Oh my God, that's disgusting. Oh, oh God. Oh.
Saito: Oh. Oh. It's bitter. It's like an ash. Oh.
Me: It has the flavour of a burnt-out candle.
Saito: It's like a bonfire.
Me: I would not have predicted that.
Saito: Oh, that is horrible.

Bitter, ashy, mouth-puckeringly astringent, it didn't taste like an apple; it tasted like leftovers from a religious ritual. Were these the chemicals behind apples' remarkable antioxidant effects? Could this flavour possibly represent something good?

It seemed a perfect summary of our current relationship with fruits and vegetables—and, indeed, all the foods we try to mine for health. We do our best to extract their secrets, isolate their compounds, measure them, and—in the case of refined and fortified products—insert them back into foods. But take out the water and the sweetness, get down to the so-called soul, and we'd likely be surprised at what's going on inside.

One thing that's undoubtedly happening when you *eat* an apple (or any other natural food, for that matter) is synergy, the phenomenon in which substances work differently when they're together than when they're on their own—the vitamin C in a Red Delicious, for example, may act very differently in isolation than it does when it's surrounded by an apple's other compounds. Or consider berberine, a chemical compound in a plant called goldenseal that's known for its antibacterial qualities. When ingested in the form of goldenseal, berberine isn't usually toxic. But the same dose of berberine "can get pretty toxic pretty quick" if it's isolated from the plant, said James Neal-Kababick, the analytical chemist who found the Viagra tablet in an "herbal" supplement. This is yet another example of why it's unwise to assume that a supplement made of an isolated natural compound is necessarily safe.

The effects of some compounds also depend on what else they're consumed with—the fat-soluble vitamins (A or beta-carotene, D, E, and K) require adequate fat to be absorbed (that's one reason cooking vegetables with oil can make them more nutritious), whereas the water-soluble vitamins (C and the Bs) do not. What's more, whole foods often appear to contain other substances that are necessary for their supposedly beneficial phytochemicals to be absorbed and used. A 2011 study on broccoli found that giving subjects fresh broccoli florets led them to absorb and metabolize seven times more of the anticancer compounds known as glucosinolates, present in broccoli and other cruciferous vegetables, than when glucosinolates were given in straight capsule form—presumably because the whole broccoli contained other compounds that helped their bodies put the anticancer chemicals to use. Indeed, many times the active forms of plant chemicals found in human blood are different from those found in food, indicating that some sort of conversion has taken place.

Tomatoes have garnered attention recently for their levels of lycopene, a red-orange carotenoid with powerful antioxidant effects—that's why lycopene has become a nutritional buzzword, popping up on labels of everything from spaghetti sauce to ketchup. (Lycopene, like all carotenoids, is fat-soluble—so it's also better absorbed if prepared with some oil.) But tomatoes contain many bioactive chemicals besides lycopene, including not just vitamins like folate and vitamins C and E, but other carotenoids like phytoene, phytofluene, and beta-carotene, and flavonols like quercetin and kaempferol. These compounds' combined presence may well explain why one 2004 paper in the *Journal of Nutrition* found that freeze-dried whole tomato powder appeared more effective than lycopene supplements alone in reducing the growth of prostate tumours in rats.

"It's not like A does B, and X gives you Y," said Neal-Kababick. "There's a lot that's still being discovered and understood. We're at the tip of the iceberg: we haven't really even gotten our heads under the water yet to see the rest of the work that has to be done."

As I'd learned back at the Combat Feeding Directorate, the intricacies of these interactions have changed the way military researchers approach nutrition: while they're studying the effects of individual phytochemicals, they've also come to the conclusion that like chemistry between people, the complexities of nutritional synergy may be impossible (or at least uneconomical) to create artificially.

"We're getting away from pure compounds—the idea of purifying a specific substance from apples and onions, for example—and moving toward extracts instead," said Betty Davis, leader of the Performance Optimization Research Team at the Combat Feeding Directorate. "It's for two reasons. First, isolating compounds is very expensive. And second, there may be issues of bioavailability and absorption and synergies that, frankly, we don't understand yet."

Not everyone is focusing on whole extracts, however. Given the profit potential of fortification, the food industry is extremely interested in plant compounds—but usually in a more isolated kind of way. Coca-Cola, for example, funded an incredibly un-Coke-like-sounding study published as "A Pilot Study on the Effect of Short-Term Consumption of a Polyphenol Rich Drink on Biomarkers of Coronary Artery Disease Defined by Urinary Proteomics"—presumably with the intention of discovering which specific phytochemicals would make sense to add to their products. And the tomato and prostate tumour study I mentioned above was originally presented at a conference on food, nutrition, and cancer that was sponsored by corporations including the Campbell Soup Company, the Cranberry Institute, the National Fisheries Institute, Hill's Pet Nutrition, and the United Soybean Board. (An educational grant was provided by the Mushroom Council.)

Though phytochemical-fortified processed foods are already available in your supermarket, it's worth remembering that the idea of isolating synergy is inherently oxymoronic—and I mean that as it sounds. As one paper on synergy put it, "Understanding one leaf in a forest does not necessarily provide insight into the entire forest. The interrelation of human physiology and of the biological activity of plant and animal foods that humans consume is incredibly complex, replete with checks, balances, and feedback loops, dependent on a myriad of substances that differ only in subtle ways from one another."

These are not the types of issues that lend themselves to examination in randomized, controlled clinical trials, the gold standard of scientific studies whose very purpose is to determine causality by examining things in isolation. Considering the thousands of chemicals in plants (and, for that matter, animal products), unravelling the secrets of their interactions with each other—and with our bodies—is a daunting, if not impossible, task.

In the meantime, Nutrilite's approach is theoretically a good one: instead of isolating every chemical on the dirty chromatogram, it tries to keep them all together—the idea being that we can reap their benefits even if we don't entirely understand how they work. But as logical as this philosophy is, it still prompts the question: Why do we need pills at all? If we really want to capture the potential health benefits of naturally occurring nutrients, why don't we just eat more foods that contain them?

Regardless of *why*, the truth is that we don't. According to the 2010 *Report of the Dietary Guidelines Advisory Committee on the Dietary Guidelines for Americans*—a 445-page, bright orange reader full of depressing revelations about our eating habits—only about 5 percent of American adults under age fifty are meeting the recommendations for dark green vegetables, and only about 25 percent of us meet the recommendations for fruits, even when you include juice (which is essentially liquid sugar). Instead, the number one contributors of calories to the American diet are "grain-based desserts" like cakes, cookies, pies, doughnuts, crisps, and cobblers.[1] If non-vitamin phytochemicals turn out to be as important as many researchers suspect, then our obsession with multivitamin pills—which don't contain these other chemicals (and whose vitamins may themselves be redundant, given how many of our foods are fortified and enriched)—could mean we're insuring ourselves against the wrong dangers. It's as if we're taking out earthquake policies in an area more at risk of floods.

Perhaps even more surprising than our potential deficiencies in non-vitamin chemicals, however, is that some of us still manage to be deficient in vitamins and minerals even *with* all of our enrichment, fortification, and multivitamin use. This may be because there are still plenty of extremely popular foods that are not enriched or fortified, such as potato chips, French fries, soda, and candy bars; the aforementioned "grain-based desserts," while often made with enriched flour, are also lacking in most other micronutrients. Adding to the problem, according to analyses of data from the

[1] What's more, these foods also consist nearly exclusively of easy-to-digest carbohydrates, which, whether sweet or starchy, are quickly absorbed into our bloodstreams in the form of glucose—that is, sugar. This causes a spike in the hormone insulin, which in turn encourages our bodies to store those calories as fat.

National Health and Nutrition Examination Survey, the people who use dietary supplements are more likely than non-users to eat diets that are naturally rich in vitamins (not to mention exercise regularly, abstain from tobacco, and drink limited amounts of alcohol). That means that most vitamins—and other nutritional supplements—are being taken by the people who need them the least.

Whatever the reason—and despite the fact that most of us consume more than enough calories—data from the National Health and Nutrition Examination Survey (NHANES) suggests many Americans are not meeting the USDA's dietary targets in micronutrients, including vitamin D, calcium, potassium, and to a lesser extent, choline, magnesium, and vitamins A, C, E, and K (not to mention dietary fibre). Blood and urine tests done by the Centers for Disease Control—which are more often accurate since they do not rely on people's memories of what they ate—also suggest that vitamins B6, D, C, and B12 deficiencies may be a concern for some people. Considering that we still haven't resolved the details of all that vitamins do in our bodies and how much we need—which raises the issue of whether the current nutritional recommendations are accurate to begin with—this leads to another question being asked by modern nutritional researchers: Could these mild to moderate micronutrient deficiencies cause problems over time?

Bruce Ames thinks so. Ames, whom I visited in his office at the Children's Hospital Oakland Research Institute (CHORI), is an acclaimed biochemist best known for his development of the Ames test, an easy and inexpensive way to test for potential carcinogens. He gained fame among environmentalists for work showing that many man-made chemicals are carcinogens, but then fell out of favour with some of them when he found that many naturally occurring compounds are, too. Born in 1928 in the Washington Heights neighbourhood of New York City, Ames is convinced that Americans' diets, which he believes are moderately low in many micronutrients, are having catastrophic health effects on our long-term health.

"My talent, such as it is, is seeing the big picture and opening up new continents," he told me as we sat across a round table in his office at CHORI, separated by a bowl of papier-mâché fruit. When he was a cancer researcher, this talent led to his recognition that DNA damage causes certain cancers, which in turn led to the development of the aforementioned Ames test.

"It got me interested in preventing disease, in figuring out what's causing DNA damage and how we can prevent it," he said. "We started growing human cells in culture, and whenever we made them short in a vitamin or a mineral, we got DNA damage—damage similar to the sort that can lead to cancer. I said, 'Hey, why the hell is nature doing that?' And then one day it hit me: it's the difference between short-term and long-term survival."

Ames calls his ensuing idea, which he first published in 2006, the triage theory, and it boils down to this: Much like a wartime medic, the body has priorities, and its first priority is to keep you alive. If a micronutrient like a vitamin or mineral is in short supply, the body will use it for its most immediate, most pressing needs first. Only after those needs are fulfilled will the leftover micronutrients become available for their other functions in the body.

Take vitamin K, which is named after the Danish word for coagulation and is known to be important for blood clotting and bone health. Ames believes it also plays roles in preventing cancers and heart disease, among other functions. Vitamin K, which is concentrated in dark green plants like spinach and Swiss chard, isn't included in most multivitamins, and it's one of the vitamins for which there's not yet an RDA.

In an ideal world, we'd have enough vitamin K for all its purposes in the body—whatever they all may be. But Ames argues that if supplies are limited—as they are in many people's diets (especially the poor, obese, and elderly)—blood clotting is going to take priority over other functions that are less evolutionarily pressing. That makes it less likely that you'll bleed to death in the short term, but it may also put you at risk of other long-term, age-related problems that vitamin K may play a role in preventing, like bone fractures and heart disease.

It's like house repair: If you've got a limited budget and you're trying to decide between preventing an immediate flood in your basement or fixing a small, non-urgent leak in your roof, the basement will probably win. That's a good decision for the short term, but the roof leak, if not addressed, could cause big problems later. The way Ames and his main collaborator on the idea, Joyce McCann, see it, the fact that many of us are eating nutritionally (if not calorically) poor diets means that our bodies simply don't have enough of the raw materials necessary to maintain our long-term health. We may still feel fine in the short term, but in Ames's view, these moderate deficiencies, when they persist over time, could cause damage that could contribute to many of our age-related problems and diseases, from cancer and heart health to osteoporosis, immune dysfunction, and dementia. Ames is specifically looking at the effects of chronic moderate deficiencies in vitamins and minerals, but the same might prove true for phytochemicals and other dietary chemicals as well.

Unlike many other nutritional researchers, Ames is a fan of daily multivitamin supplements, particularly for the poor, and has spent several years developing something called a CHORI-Bar—which is essentially a multivitamin in the form of a dense black, vaguely chocolate-flavoured brick, sprinkled with sugar crystals, that he hopes might help raise the nutritional status of people who aren't taking multivitamins, or who can't afford (or don't want to eat) more naturally nutrient-packed diets. But regardless of the means, he's convinced we have to do *something.*

"I have this vision that the big thing that's causing ill health is what we're doing to *ourselves,*" he told me. "The low-hanging fruit in preventive medicine is nutrition, and that's where we have to put our efforts. I'm eighty-three years old, and I don't know how many years I have left to communicate this."

Regardless of whether the triage theory is correct, Ames's work touches upon another enormous new area of nutritional research. That is, nutritional genomics (sometimes just called nutrigenomics), the emerging study of how our genes determine how our bodies react to food—particularly in terms of the development of disease—and how our food might actually affect our genes.

In theory, the emerging field of nutritional genomics could enable us to create truly personalized diets, designed around each person's genetically determined needs and sensitivities. And indeed, a number of companies already claim to be able to offer personalized diet recommendations based on genetic idiosyncrasies, including the weight-management genetic test offered by a company called Inherent Health. According to Inherent Health's website, the test "take[s] the guesswork out of losing weight" by analyzing specific portions of your DNA.

"Don't waste another day on the wrong diet!" its copy proclaims. "What's the right percentage of carbs, fat and protein for your diet? Should you exercise at high intensity, or can you get by with a moderate workout? Your test results may tell you the personalized answers you need to achieve your healthy weight loss goals."

Most scientific organizations are extremely dubious about the value of today's diet-oriented genetic tests, though many scientists believe that they could be useful in the future. But before getting too deep into their usefulness, let's devote a moment to the science they rely on. As you likely recall from high school biology, genes (from the Greek word *genos*, meaning "birth") are the snippets of code, wrapped in the double helix of our DNA, that pass down inherited traits from one generation to the next. Often described as being like sentences, genes are each made from a long series of genetic "letters" from a four-letter alphabet—A, T, C, and G—in which each letter represents a nucleic acid base (adenine, thymine [not to be confused with thiamin], cytosine, and guanine). Each of these genetic "sentences" contains the instructions for how to make biological products, the best understood of which are proteins.

Proteins are large molecules that are responsible both for helping our bodies *do* stuff (hormones and enzymes are proteins, for example) and for creating much of the structure of our tissues themselves. As a result, genes—or, more precisely, the proteins they create—are what determine inherited traits like your hair colour or which hand you write with. According to the latest estimates from the Human Genome Project, humans have about 25,000 protein-coding genes.

The market for genetic tests that are sold directly to consumers—whether they're meant to determine your ideal diet, your baby's father, or your predisposition for particular diseases—began to blossom after the successful mapping of the human genome in 2003. The tests search for genetic variations called single nucleotide polymorphisms, usually referred to as SNPs (pronounced "snips"), which are spots on our DNA when one genetic letter has been replaced with another. SNPs, which account for most genetic variation among people, occur roughly once every three hundred genetic "letters" and are usually found in areas of DNA that don't code for proteins (only about 3–5 percent of our DNA codes for proteins; the rest is still largely a mystery).

Our DNA contains roughly three billion base letters, which means every person has about ten million SNPs. Since most SNPs are caused by replication errors (the genetic equivalent of typos), and because we need to create a full copy of our DNA for every new cell in our body, we can accumulate additional SNPs over our lifetimes—some of which we then pass down to our children. New SNPs can also occur between generations: every baby has about fifty to seventy new SNPs, created in large part by pure chance.

To reiterate, SNPs aren't actually genes; they're just individual swapped letters within genes. As such, SNPs don't normally seem to matter—if I wrote "ladddr" instead of "ladder," for example, you'd probably be able to figure out from context what I meant. But writing "dick" instead of "duck" would change your subject matter entirely. Likewise, some SNPs alter genes' codes in a way that either changes their expression (that's the scientific way of describing whether they're turned on or off) or changes what proteins they create. Some of these changes may have neutral or even protective effects— in fact, beneficial SNPs are essential for positive evolution, since they can lead to useful traits like an ability to survive at higher altitudes. But others are thought to affect gene expression in a way that can cause disease. As the National Coalition for Health Professional Education in Genetics explains on its website, "Because genes and gene products are involved in all disease processes, the question should not be, 'Is this a genetic disorder?' but rather, 'What role do genes play in the expression of this disease in this person?'"

Both our genes (the sentences) and our SNPs (the typos) can determine our sensitivity to particular environmental factors, including foods and drugs. This might explain why certain drugs work better for some people than for others, why many nutritional studies appear to have contradictory results (for example, some studies have found that caffeine increases your risk of bone loss, whereas others deem it actually beneficial), why certain populations are more susceptible than others to particular diseases, like type 1 and type 2 diabetes, and why disease risk may vary even among people whose lifestyles and diets are essentially the same.

As our understanding of the relationship between our environment and our genetics grows, we may eventually be able to use genetic tests to make dietary recommendations down to the individual (or quasi-individual) level. Theoretically, this could clear up confusion over what to eat by helping people determine which nutritional advice to follow and which they can safely ignore. Perhaps a genetic test would reveal that you can eat higher amounts of carbohydrate than other people without gaining weight. Or maybe it would show that your vitamin requirements are higher than most of the population, and that you therefore should take a supplement. The more sophisticated the tests are, the more personalized our diets could become.

And to push this idea to the extreme, if you combined this type of genetic profiling with accurate, early, fast tests for nutritional deficiencies, there wouldn't be a need for population-based nutritional guidelines like the RDA at all. As Gerald Combs writes in his textbook *The Vitamins*, "The time is quickly approaching when it will be possible to identify disease predisposition, metabolic characteristics, and specific dietary needs of individuals based on rapid, genomic/metabolic analyses. As that becomes practicable, the population-based paradigm will lose much of its value."

But despite the promises of consumer-oriented genetic testing companies, we're not there yet—in large part because, while we can identify genetic SNPs with surprising precision, we don't yet know what they actually mean, how they might work together or influence each other, or whether we're even looking for the right ones. (This is part of the reason that the FDA has prohibited the sale of some of the tests.) Consider this analogy given by José Ordovas, PhD, director of the Nutritional and Genomics Laboratory at the Jean Mayer USDA Human Nutrition Research Center on Aging at Tufts University, who's been working on the subject of personalized nutrition for more than twenty-five years.

"It's like someone rings your doorbell and you have one of those peepholes to look through. Well, who's coming? First of all, the guy might hide so you don't see *anything*. Or you could see that particular guy, but not the twenty others who are standing on either side of him. You're looking at what's directly in front of you, but you're missing everything else that's happening outside of your vision."

As a result, at least for the time being, it doesn't make sense to fork over hundreds of dollars for consumer-oriented dietary genetic tests. If the ones I've personally tried are any indication, they'll likely just parrot back the same nutritional advice (eat more produce and fewer processed foods) that you were hoping the test would help you avoid having to follow.

As scientists work toward this goal of creating personalized (or semi-personalized) nutritional advice, every step we take seems to lead to another mystery—including the other side of nutritional genomics: If your genes can determine how your body responds to food, then what influence might your food have on your genes? Like so much about nutrition, this is an open question. But what we know so far is pretty crazy.

The question of what effect our food can have on our genes is often brought up in the context of a World War II tragedy, known as the Dutch Hunger Winter, that began in September 1944. Allied forces had launched a parachute drop behind the Nazis' front lines near the Dutch city of Arnhem in a bold attempt to clear the way for a land invasion of Germany, but the operation was a failure, and the Nazis retaliated against the Dutch resistance's support for the Allies by declaring a total embargo on occupied Holland.

The cities in the western part of the Netherlands were already short on food, and the Nazis relented slightly, allowing supplies to be transported by water. But it wasn't much help: the winter of 1944 came early, and the canals quickly froze, preventing supplies from being delivered by barge. Once that happened, the famine began in earnest. The official daily rations per person allocated by the Germans dropped from an already low 1,400 calories in November 1944 to 400–800 calories from December through the following April. Though people tried to scavenge calories from everything including grass and tulip bulbs, nearly all of their energy came from three foods: bread, potatoes, and sugar beets. The famine reached its peak in April 1945, when the Allied advance completely cut off the western cities from the rest of the country. By the time the Netherlands was liberated in May—at which point the normal food supply was quickly reestablished—much of the population of its western cities had been starving for months.

The Dutch Hunger Winter killed at least 22,000 people and left another 200,000 sick from starvation; the average weight loss among surviving citizens was 15–20 percent. By the famine's end, around half the women had stopped menstruating, and nine months after the worst months of the famine, the birth rate dropped to less than 50 percent of its previous level—a reflection of the effect the most acute period of starvation had on fertility.

It was an avoidable tragedy, caused by military strategy and human politics. But the Dutch Hunger Winter also created a very interesting research opportunity, a situation known as a "natural experiment," in which a historical event—usually a bad one—results in conditions that would be logistically and morally impossible to create deliberately. Not only had the Dutch Hunger Winter affected a huge number of people for the same, clearly defined period of time, but it had cut across all socioeconomic groups. The affected people had all eaten similar foods; what's more, only the urban west of the country had endured the famine, providing a culturally matched non-starved control group. The famine had also occurred in a country known for its meticulous record keeping, giving later researchers access to decades' worth of detailed, ongoing health data from medical and military exams. The Dutch Hunger Winter therefore created a tremendous opportunity to examine connections between nutrition and health—and, in particular, how malnourishment in pregnant women might affect their babies.

When these studies began in the 1970s, it was already known that what people ate could have an immediate effect on their body chemistry. We've seen this with vitamin deficiency diseases: Our bodies need vitamins for particular chemical reactions. If we don't have enough of them, those reactions can't take place—and we get sick. Similarly, environmental toxins like airborne asbestos have been associated with negative health effects like lung cancer. But until recently, the assumption was that while environmental factors can cause the expression of disease-causing genes, their effects stopped with the person who experienced them; in most cases, they wouldn't be passed on to the next generation. DNA itself—at least in the short term—was supposed to be immutable.

This attitude began to change in the second half of the twentieth century, when research showed that factors like smoking cigarettes while pregnant did in fact have negative health effects on babies. But the results of the Dutch Hunger Winter studies took this idea a step further, focusing particularly on nutrition, not toxins. Depending on when in their mothers' pregnancies the famine had occurred, the children of the malnourished women were at increased risk of later physical and mental health problems as adults, including depression and cardiovascular and metabolic diseases like type 2 diabetes. (Similar effects have recently been seen in Gambian children, depending on whether they were conceived during the rainy or dry season.) Children conceived during the peak of the Dutch famine also had a twofold risk of developing schizophrenia compared with those conceived at different times, a connection supported by a similar natural experiment: the horrendous Chinese famine that resulted from Mao's so-called Great Leap Forward.[2]

We've all heard of the nature versus nurture debate—the argument over whether our genes or our environment are the true arbiters of our disease risk, intelligence, personalities, and all the other factors that make individuals unique. In this case, as is true for babies whose mothers smoked during pregnancy, it might seem easy to argue that the babies of the Dutch Hunger Winter simply suffered from bad embryonic nurturing: their higher incidence of disease was the consequence of having depended on the bodies of their nutritionally depleted (and probably pretty stressed out) mothers for food. According to this theory, the babies were somehow influenced by their embryonic environments, but not in a way they'd pass down to their own children.

But here's the shocking thing: these effects *did* appear to get passed down. Subsequent research has suggested that exposure to the famine may have affected the mental and physical health of the mothers' *grandchildren* as well. For example, babies of women whose *mothers* had been exposed to the famine while in utero have been found to have higher levels of body fat at birth and to have poorer health later in life than people whose mothers' mothers did not suffer the famine while pregnant. (Also, grandchildren of smokers appear to be at higher risk of asthma.)

As one academic article put it, "We are only beginning to appreciate the generation-spanning effects of poor environmental conditions during early life, which may be particularly relevant to populations in transition between traditional and western lifestyles. This may shed light on the epidemic of diabetes, obesity and cardiovascular disease, which is rapidly expanding in

---

[2] The reason for the higher rates of schizophrenia among the famished mothers' children is still under debate, but at the moment the strongest dietary contender is folate, partially because the peak in risk for schizophrenia coincided with a peak in central nervous system abnormalities (neural tube defects), which are known to be linked with folate deficiencies. Interestingly, "Association Between Maternal Use of Folic Acid Supplements and Risk of Autism Spectrum Disorders in Children," a 2013 paper published in the *Journal of the American Medical Association*, also found an association between prenatal folic acid supplementation around the time of conception and a reduced risk of autism in offspring. If this association were proved to be causal (which it has not!), it adds a new angle to the investigation of the effects of folic acid–enriched grain products in the United States, which became mandatory in 1998.

such countries"—all of which, in addition to osteoporosis, neurological disorders, and a variety of inflammatory conditions, are thought to have connections to diet.

This transgenerational effect, as they call it in the research literature, would seem to knock out the strict nurture hypothesis, since the grandbabies were never directly exposed to their grandmothers' wombs. What's more, other research suggests that some disease risk can be passed down to grandchildren via their grand*fathers*, who obviously do not have wombs in which to do any nurturing. But the transgenerational effects were unlikely to have been caused by changes to the babies' and grandbabies' genetic nature, either, since population-wide changes in genes themselves usually take far longer to develop. In short, the children's and grandchildren's health risks did not appear to be caused purely by their genetic nature *or* their prenatal nurturing. Instead, they seemed to have to do with how the babies' nature itself was nurtured.

My first personal exposure to this idea of nurtured nature came my senior year of college when I was diagnosed with type 1 diabetes, an incurable autoimmune disease in which the body destroys the cells that make an essential hormone called insulin. As I adjusted to life with the disease, I wanted to know: Had developing type 1 diabetes been my genetic fate? If so, why hadn't I done so until I was twenty-two years old? Instead of an explanation, doctors kept offering a metaphor. "Genetics loads the gun," they told me. "But it's your environment that pulls the trigger."

As someone who can be excessively literal, I did not find this to be a particularly satisfying response. But it wasn't my doctors' fault. The question of how environmental factors (including but not limited to diet) affect the activity of our genes—let alone how these changes may affect future generations—is an extremely new area of scientific research.

It's known as epigenetics, a term that refers to a secondary set of instructions that tells our genes where, when, and to what extent they should be expressed. (The word "epigenetics" means "above" or "in addition to" genetics.) To borrow an analogy from British scientist Nessa Carey, author of the excellent book *The Epigenetics Revolution*, you can think of this in terms of a Hollywood movie: if DNA is a script, epigenetics are directions that determine how it'll be produced. For example, the 1996 adaptation of *Romeo and*

Juliet starring Claire Danes and Leonardo DiCaprio was based on Shake-speare's script, but was quite a different production from the version origi-nally put on at the Globe Theatre. These specific instructions are referred to as epigenetic "marks," and can be created by external factors like chronic stress, emotional trauma, exposure to chemicals, and yes, food.

We are surrounded by examples of epigenetics, even though we may not recognize them as such. Think about it: every cell in your body (with the exception of red blood cells) contains an entire copy of your DNA and, with it, every one of those 25,000 or so protein-coding genes. DNA in your heart cells contains information on how to build an eye; DNA in your intestines also knows that you're right-handed. And yet, thanks to epigenetic marks, our feet know not to grow teeth, and our stomachs don't grow ears. Women's bodies know not to make breast milk all the time. What's more, depending on which genes are turned on and off, the same creature can look completely different at different stages of its life—think of a caterpillar that morphs into a butterfly.

Epigenetic marks don't actually change the underlying genes themselves (and they are different from SNPs, which are actually incorporated into our DNA); they just tell genes how to behave. As such, epigenetic marks can be permanent or temporary, which explains why a woman only produces breast milk during certain times of life, but a butterfly will never revert to a caterpil-lar. If you think of DNA as being like a computer's hard drive, packed with programs, epigenetics is what decides how and when they should be run. Or, to bring it back to the diabetes example, if genetics loaded the gun, epige-netic factors—which my doctors called the "environment"—pulled the trigger.

Cells pass down many of their epigenetic marks every time they divide—that's why when a muscle cell replicates, the new daughter cells know that they're supposed to be muscle cells, not liver cells. However, most epigenetic marks are not transferred to our actual kids; instead, the body performs a reprogramming at the time of conception that appears to remove most of the epigenetic marks a zygote's parents have accumulated in their lives. This blank slate allows the union of a single sperm and egg to differentiate into a brand-new human body. But most is not all. Some epigenetic marks do appear to be heritable. And besides, even the most thorough reprogramming would still leave a fetus susceptible to new epigenetic marks added during its time in the womb.

In the Dutch Hunger Winter studies, both of these factors seem to be in play. In the first generation—that is, the children of mothers who were directly affected by the famine—epigenetic signals caused by the pregnant women's starvation seem to have permanently affected the expression of their babies' genes. Those babies then appear to have passed some version of those acquired epigenetic marks on to their own children.

Let's pause here for a moment to acknowledge how heretical this idea is. Read enough about epigenetics and you'll inevitably come across a reference to Jean-Baptiste Lamarck, a pre-Darwinian French scientist whose 1809 book, *Philosophie Zoologique*, introduced the theory of acquired traits, which proposed that species could acquire new characteristics from environmental influences and pass them down to their offspring. Blacksmiths pass their hard-earned brawny forearms down to their sons, Lamarck argued, and a giraffe, whose neck Lamarck believed had been elongated through constant stretching for high leaves, would give birth to a similarly elegant calf.

Fifty years later, Charles Darwin published *On the Origin of Species*, and the success of his theory of evolution pushed Lamarckism into the scientific dustbin. For much of recent history, Lamarckism has been considered preposterous, a remnant from an unsophisticated scientific era. Common sense now holds that if blacksmiths' sons develop brawny forearms, it's because they did some sort of physical activity that built up their muscles, and whatever French I acquired in school is not going to be directly transferred directly to my child. (*Tant pis!*)

But amazingly, the rapidly developing field of epigenetics is suggesting that Lamarck may have been on to something—not in terms of acquired *knowledge* (those French verbs will die with me) but perhaps in terms of other acquired traits. As one 2012 academic review put it, "The finding that the establishment of the epigenome can be influenced by environment, in combination with the finding that a few epigenetic marks escape reprogramming between generations, raises the possibility that environmentally induced epigenetic marks could be inherited by the next generation.

"If this were true," the authors continued, "it would profoundly change our understanding of inheritance."

The nutritional implications of this idea are particularly important, since diet is an area over which individuals have a considerable amount of control. It suggests that we aren't just what we eat, or even what our food eats. We're what our grandparents and parents ate. And our grandchildren may be affected by what *we* eat, too.

While this aspect of nutritional genomics—the idea that our diets could have transgenerational epigenetic effects—is still controversial, a highly publicized 2003 paper from researchers at Duke University demonstrated the potential vividly. Their experiment is frequently illustrated by a photograph of two mice, the modern-day equivalent of the photographs of lab rats that the agricultural chemist Elmer McCollum used to carry around in his pocket to show the importance of vitamins. In the picture, one mouse is brown, lean, and unremarkable, the same type of mouse you may have seen scurrying in your basement. The other mouse is yellow and enormously obese, the rodent equivalent of a beach ball. Unless you've grossly overfed a pet hamster, you've never seen a mouse like this. The mice are unmistakably different. But they're also identical twins.

Their physical differences are caused by what's known as the *agouti* gene— hence their band-like nickname, the Agouti Sisters. The *agouti* gene (which humans don't have) affects fur colour, thus providing a convenient visual clue for scientists trying to determine how strongly it's been activated: if it's off, the mouse is brown, and if it's on, the mouse is yellow (if it's only partially expressed, the mouse is mottled). The *agouti* gene also causes an imbalance in hormones that results in insatiable hunger, which leads to obesity and puts the mouse at risk of type 2 diabetes and cancer. Just from looking at the photograph, it's clear that the obese yellow mouse's *agouti* genes are strongly turned on.

There are likely many ways by which epigenetic factors can turn genes on and off—and we don't know all their details—but the *agouti* gene is thought to be related to the process we understand the best: methylation. Methyl groups are basic structures in organic chemistry that consist of one carbon atom attached to three hydrogen atoms. If methyl groups bind to certain spots on DNA or the proteins around which DNA is wrapped, the gene controlled by that section of DNA will be affected. In the case of the mice, the brown mouse's *agouti* gene had been methylated (which had turned it off), whereas

that of the yellow mouse had not been methylated, and therefore had been
turned on.

In this particular experiment, the researchers were interested in finding
out whether they could influence methylation patterns—and thus which genes
were turned off or on—by changing what the mice ate. First, they fed pregnant
mice whose *agouti* genes were not expressed (and who therefore were lean and
brown) a diet that included a chemical called bisphenol-A, more commonly
known as BPA, which is used in plastic products like water bottles, the tops of
disposable coffee cups, the linings of food cans, baby bottles, dental sealants,
and even in the ink that is used to print receipts.

BPA has become controversial in recent years in part because of its potential effects on gene expression, and at least in the case of the mice, this theory seemed to hold true: the DNA of the babies born to BPA-exposed
mothers—including the section of their DNA that contained the *agouti*
gene—was less methylated than that of the mice whose mothers had not
consumed it; presumably as a result of their exposure to the dietary BPA,
their *agouti* genes had remained on. This was immediately visually obvious.
Whereas the mothers were lean and brown, most of their babies were obese
and yellow.

Then the researchers performed a second experiment. In addition to BPA,
they gave pregnant brown mice supplements of folic acid, vitamin B12, choline, and betaine, a chemical that's found naturally in seafood, spinach, beets,
and wine. All of these substances contain methyl groups that the body can
use for DNA methylation, making them what's known as methyl donors. When
these supplemented mice gave birth, most of their babies were lean and brown,
even though they'd been exposed to BPA. The dietary supplements—three
of which were vitamins (if you count choline)—seemed to have provided the
methyl groups necessary to keep the *agouti* genes turned off. In the Dutch
Hunger Winter schizophrenia studies, supporters of the folate theory suspect
that something similar happened in humans: a folate deficiency at the time of
conception may have affected the methylation—and therefore expression—
of certain genes in a way that eventually led to psychiatric disorders in the
mothers' offspring.

While nutritional genomics and epigenetics are both extremely new areas
of research—and it's never safe to assume that something that happens in a
mouse also happens in humans—examples like this do strongly push toward

an intriguing and disturbing conclusion: that our daily decisions about what to eat (and our exposure to environmental chemicals) may affect the expression of our genes.

As interesting as the *agouti* mice research is, it doesn't mean that we should all start popping supplements—on the contrary, our understanding of how diet affects an embryo is embryonic itself. The Duke researchers stressed that just because dietary-induced methylation seemed to be beneficial in one gene does not mean it does good things in all others, or that their findings in mice should automatically be transferred to humans. As they put it (in reference to the addition of folic acid to enriched grain products, which became mandatory in the United States in 1998), "Population-based supplementation with folic acid, intended to reduce the incidence of neural tube defects, may have unintended influences on the establishment of epigenetic gene-regulatory mechanisms during human embryotic development." Translation: We don't know yet what else it might do.

Nor do we know what the long-term epigenetic effects are of eating so few natural foods like vegetables (and their accompanying phytochemicals) or so many refined and processed products. We don't know the consequences of our national taste for fad diets, or our habit of swinging wildly from one eating pattern to another.

And we certainly don't know our diets' effects on our microbiomes, the countless bacteria, viruses, fungi, and protozoa that inhabit our intestines and appear to play an intimate and important role in our health. The National Academy of Sciences has called the microbiome "arguably the most intimate connection that humans have with their external environment, mostly through diet," and its population is nearly unbelievably large. The Human Microbiome Project estimates that microbial cells outnumber our human cells by a factor of ten to one; even though each microbial cell is only one-tenth to one-hundredth the size of a human cell, our combined microbiome is estimated to weigh between 1 and 2 percent of our body weight. Whereas the human body has some 25,000 protein-coding genes, our microbiome is estimated to have about 3.3 *million* of them, and each gene's expression could theoretically be influenced by what we eat. What's more, many microorganisms produce substances themselves that complement our diets (for example, some intestinal bacteria can make vitamin K). The number of unanswered questions is mind-boggling.

For now, it might seem strange to end a book about vitamins with such exotic-sounding subjects as phytochemicals, synergy, nutritional genomics, and the microbiome—subjects that appear worlds away from scurvy-stricken sailors and pellagra-afflicted prisoners. But swap any of these modern terms with "vitamins" and you'll see that our situation today is surprisingly parallel to that of our turn-of-the-twentieth-century predecessors. Like them, we are uncovering holes in our understanding of nutrition. Like them, we are becoming aware of the potential dangers of our arrogance. And like them, we are at the cusp of discovering entirely new ways that what we eat might affect our health.

Whether it's Elmer McCollum trying to identify what in milk was protecting his rats, or scientists experimenting with *agouti* mice, these stories are all different spots on the same historical continuum, each representing the cutting edge of science in its time. It seems likely that, just as early nutritional researchers couldn't imagine some of the things (like vitamins) that we take for granted, a hundred years from now people may marvel at our hubris in thinking we have nutrition figured out.

But vitamins, more so than any component in food, aren't just a cautionary tale. They also teach us about ourselves—about our hopes, about our fears, and about our desperate desire for control. Rather than ask questions that might be unanswerable—or challenge the food industry status quo—we passively accept whatever new health claim or recommendation we hear. Indeed, we seem to *want* to do so: it's comforting to think that even if we ourselves find nutrition confusing, there's someone out there who knows the truth.

As a result, we continue to accept the idea that anything that contains vitamins must be good, despite the fact that we viscerally know that marketers are using this assumption to manipulate us into buying their products. We don't ask where the synthetic vitamins in these foods come from, or why our food supply requires so much reverse engineering to begin with. Instead, we allow our capacity for rational thought to be hijacked by a word. And despite the fact that more than half of us take vitamins as pills (and nearly all of us associate them with health), nearly *none* of us stop to wonder why—out of all of the thousands of chemicals in food—we revere these particular thirteen, why we regard them not just with appreciation, but with what often resembles religious faith.

That's perhaps the ultimate question to be asked about our relationship with vitamins, and I've come to believe that its answer lies in the very reason faith exists: it is a salve against uncertainty. Humans hate the unknown. We chafe against it; it makes us feel powerless and paralyzed. So we assign names to chemicals; we count calories and classify food types; we look for advice on food labels and in the news; we do whatever we can to maintain a sense of control over our bodies and the world. In a high-stakes situation like health, where explanations often are incomplete and guarantees are impossible, we soothe our discomfort by finding something to believe in, something that will make us feel safe. In the case of religion, we put our faith in gods. And in nutrition, we have vitamins.

Epilogue

Uncertainty can be uncomfortable, especially when it comes to food—a subject that we literally ingest. But while this sense of responsibility can feel paralyzing, it can also be empowering; it just needs to be faced head-on. In nutrition, this means acknowledging that we have still not figured out everything about how food interacts with our bodies, and then using that very fact to decide what to eat.

Consider the example of epigenetics. If your diet today might affect your child or your grandchild—but no one can tell you exactly how—then what are you supposed to have for dinner tonight? If that question sounds overwhelming, I suggest adopting the viewpoint of Nessa Carey, author of *The Epigenetics Revolution*. "We are complex organisms, and our health and life expectancy are influenced by our genome, our epigenome, and our environment," she writes. "But remember that even in the inbred *agouti* mice, kept under standardized conditions, researchers couldn't predict exactly how yellow or how fat an individual mouse in a newborn litter would become. Why not do everything we can to improve our chances of a healthy and long life? And if we are planning to have children, don't we want to do whatever we can to nudge them that bit closer to good health?"

In other words, why not accept that we don't have all the answers—and then use what we *do* know to stack the odds in our favour?

Viewed in this light, our understanding of vitamins can serve as a useful guide, even though—or perhaps, precisely because—it is incomplete. We know that our bodies need vitamins, and that without them, we would die.

We know that the more heavily processed and refined a food is, the fewer of its original vitamins (and other potentially important chemicals) are likely to remain—and the more it will need to be enriched and fortified to make up for what's been lost. We know that superdoses of vitamins have not been shown to be helpful, and in some cases may in fact be harmful. We know that our psychological attachment to vitamins makes them powerful marketing tools, and is often used to lure us into buying foods that, were it not for their vitamins, we might—and should—otherwise reject. We know that if we rely solely on multivitamins or fortified products, we will miss out on whatever other chemicals naturally exist in foods, many of which appear to be important for our health. And, on the flip side, we know that if a food naturally contains a lot of vitamins, it likely contains other beneficial compounds, too.

With these facts in mind, deciding what to eat from a micronutrient perspective becomes easy. Choose foods that are high in vitamins that nature—not humans—put there; chances are that they're nutritious in other ways as well. (Indeed, if you focus your purchases around minimally processed foods, you don't need to explicitly shop for vitamins at all.) Before you buy an enriched or fortified product, stop and ask yourself, "*Why* has it been fortified? Would I buy it if it hadn't been?" If not, substitute a food that is naturally nutrient-dense. Instead of asking if you should take a multivitamin, ask yourself about the quality of the food that you're eating—and try to improve your diet before turning to a pill. And if you genuinely find vitamin-packed energy bars and other artificially fortified products to be delicious, then fine, indulge once in a while. But acknowledge that you're actually eating a candy bar sprinkled with a multivitamin; don't let its fortification give it a nutritional free pass.

If we followed this advice, we likely would end up with a modern version of the Protective Diet that the vitamin researcher Elmer McCollum recommended to his *McCall's* readers nearly a hundred years ago, an approach that emphasized eating a wide range of naturally nutrient-dense foods as a way to hedge your bets. Based on a similar philosophy, today's Protective Diet would ask us to judge foods' nutritional value not by what has been added to them, but by what micronutrients their unfortified selves contained. It would also encourage us to deliberately choose foods with the highest natural concentration of nutritional unknowns, casting the net wide for compounds whose importance we don't yet understand. This would steer us toward fruits and vegetables, whole grains eaten truly whole—not transformed into flour or crackers or cereal—and nuts. It would include minimally processed meat, especially fatty fish, and dairy products, too. It would emphasize raw produce, but it would also recognize that there are many cases, in particular those involving fat-soluble vitamins, where cooking can increase the availability of micronutrients.[1] The best advice, from a micronutrient perspective, is to just eat a lot of vegetables, fruits, and other naturally nutrient-dense foods, regardless of how they're prepared.

Today's Protective Diet could also be defined not just by what it would include—much of which, frankly, is stuff we all know we should be eating anyway—but by what it would reject: namely, refined and processed grain products, sugary beverages, and any other attempt by humans to reverse engineer food. By encouraging the consumption of naturally nutrient-dense foods, it would also rectify a paradox in our approach to nutrition: for as obsessed as we are with the idea of nutritional recommendations, we don't follow the ones that we already have.

And last, by freeing us from the need to micromanage, by encouraging us to embrace the unknown rather than obsess over it, the Protective Diet would provide a buffer against the constant barrage of conflicting nutritional advice that we encounter in the media. We'd be able to remember that the need to generate "news" often trumps responsible reporting, and that most scientific breakthroughs don't occur in a day. If we follow a modern Protective Diet, we don't need to chase headlines or turn our meals into math problems. While it may seem counterintuitive, the very simplicity of the concept makes it the most scientific—and, I would argue, enjoyable—way to eat.

---

[1] For example, a 2002 study in the *European Journal of Clinical Nutrition* found that only 3 percent of total beta-carotene in carrots was released when they were eaten raw, versus 21 percent when they were blended to a pulp, 27 percent when that pulp was cooked, and 39 percent when the pulp was cooked with oil.

ACKNOWLEDGMENTS

I am extremely grateful to my agent, Jay Mandel, for seeing promise in my original proposal, and to Ann Godoff and Benjamin Platt at the Penguin Press for guiding that proposal into this book.

The support of the Alfred P. Sloan Foundation's Public Understanding of Science, Technology and Economics program made it possible for me to immerse myself completely in this project. The Société de Chimie Industrielle and its American section fellowship introduced me to the historical resources and lively community of the Chemical Heritage Foundation, and the Mesa Refuge at Point Reyes Station provided mental and physical space in which to polish the first draft. My experiences at the Food and the Medical Evidence Boot Camps for Journalists at the Knight Science Journalism Program at MIT, as well as the Mary Frances Picciano Dietary Supplement Research Practicum at the Office of Dietary Supplements at the National Institutes of Health, greatly enhanced my background knowledge.

I could not have written this book without the help of the many people who took time out of their busy schedules to answer my questions, tell me their stories, share their opinions, and point me toward additional resources. They include Al Sommer, Gerald Combs Sr., Bruce Ames, Tod Cooperman, James Neal-Kababick, Daniel Fabricant, David Kessler, Angela Pope, Jean-Claude Tritsch, Steve Mister, José Ordovas, Ingo Potrykus, Peter Beyer, Salim Al-Babili, Michael Levine, and Takeshi Saito. A special thanks to Suzanne Junod, Donna Porter, Nessa Carey, the FDA's communications department, Gerald Combs Jr., and James McClung for taking personal time to review relevant portions of the manuscript for accuracy—any remaining errors are my own.

I also received help from many people whose names do not appear in the finished text, including Andrea Martin and Michael McBurney at DSM, Ashley Augustyniak at the Chemical Heritage Foundation, Guy Crosby at America's Test Kitchen, David Accetta and Jeremy Whitsitt at the Natick Army Labs, A. Catharine Ross at Pennsylvania State University, Linda Meyers of the Institute of Medicine's Food and Nutrition Board, Alanna Moshfegh, Meghan Adler, and Pamela Pehrsson at USDA, Christine Pfeiffer and Rosemary Schleicher at the Centers for Disease Control and Prevention, Paul Coates and Regan Bailey at NIH's Office of Dietary Supplements, Chelsey Fields at Burpee, Lindsay Pott, Keith Randolph, J. Peter Debus, Kevin Gellenbeck, Barry Traband, and W. Kip Johnson at Nutrilite, Marjorie McCullough at the American Cancer Society, Gene Lester, John Stommel, and Janet Slovin at the USDA's Agricultural Research Service, Linda Lakind at Firmenich, and Raymond Rodriguez at UC Davis's Center of Excellence for Nutritional Genomics. Thank you as well to Jeanette Gingold, Sharon Gonzalez, Sarah Hutson, and Brooke Parsons for helping to copyedit, produce, and publicize this book.

While I came to critical conclusions about Americans' obsession with vitamins and the dietary habits that synthetic vitamins have enabled, it's important to emphasize that inadequate vitamins (and minerals) continue to threaten the lives and productivity of millions of people around the world. I am grateful to people like Klaus Kraemer and organizations like Sight and Life for their work bringing essential micronutrients to those in need.

Many friends and family members contributed emotional and editorial help, including James Vlahos, Nathanael Johnson, Jennifer Kahn, Marie Dalby Szuts, Eleanor Johnson, Jim and Hannah Leckman, Steve Korovesis, Mark Hatzenbuehler, Miriam Stewart, Michael Zimmer, Kristina and Loren Kittilsen, Adam Benforado, Brooke Bailer, Natalie Kittner, Jessica Apple, Michael Aviad, Cynthia Gorney, Deirdre English, Brigitte Bentele, Maxine McClintock, Nina Newby, Todd Rice, Betty Roche, Al Hanssen, Bonnie Hamilton, and Mary Roach.

I am especially grateful to Josh Berezin, who may want to consider becoming a professional editor if the whole psychiatry thing doesn't work out, and to Michael Pollan, who has been a source of inspiration and encouragement for more than ten years. It's difficult to convey how much I appreciate his thoughtfulness, kindness, and generosity, and the influence he has had upon my work. Likewise, words cannot capture my gratitude for Vanessa Gregory, whose friendship, wisdom, humour, editorial insight, and willingness to read the manuscript multiple times, often on short notice and on vacation, have both kept me sane and made this book infinitely better. I cannot wait to return the favour.

I would never have pursued a career as a writer if it were not for my parents, whose encouragement and faith in me has never wavered, and my grandmother, whom I dearly miss. And last, but never least, is Peter, who came up with the idea for this book and has been living with the consequences ever since. You are my most essential nutrient. I love you so very, very much.

APPENDIX A: THE VITAMINS

VITAMIN A (RETINOL): *PROPOSED: 1915, ISOLATED: 1937, STRUCTURE DETERMINED: 1942, SYNTHESIS ACHIEVED: 1947*

Called retinol because of its essential role in the function of the retina in the eye, vitamin A from dietary sources is found mostly in animal products like liver, oily fish, egg yolks, and dairy products like whole milk and cheese; you can also get it from foods that have been fortified with vitamin A like skim or reduced-fat milk, margarine, and some breads and cereals. Your body can make the active form of vitamin A from several of the plant chemicals known as carotenoids, particularly a red-orange pigment called beta-carotene that creates the bright colours of fruits and vegetables like carrots, cantaloupe, apricots, and sweet potatoes. Beta-carotene is also found in dark green leafy vegetables like kale and spinach, but its colour is masked by chlorophyll, the chemical that makes plants green.

Vitamin A is usually considered the riskiest of the vitamins because it can be toxic at relatively low doses and, since it's fat-soluble, builds up in body tissues and is hard to excrete. Usually it's difficult to reach those doses through nonfortified or enriched foods, but there are a few notable exceptions: certain animals, including seals, polar bears, halibut, and huskies, have excessively high amounts of vitamin A in their livers. And I mean excessively high—whereas humans store about 300 IU per gram in their livers, polar bear livers contain about 20,000 IU per gram. To put this in context, the Tolerable Upper Intake Level for adults—above which adverse effects are likely to occur—is 10,000 IU per *day*.

Early explorers did not know this. So when the dogsled carrying his expedition's food fell into a crevasse, the Australian Antarctic explorer Douglas Mawson and his surviving teammate, a Swiss mountaineer named Xavier Guillaume Mertz, attempted to survive by eating their sled dogs—including, of course, their livers. Several weeks later, the skin of both men was peeling off their bodies, along with clumps of hair. In his book *Polar Journeys*, Robert Feeney writes the following description: "Mawson remembered Mertz stating, 'Just a moment,' then reaching over and lifting from his ear a perfect skin cast. Mawson did the same for him, and there was hair and skin throughout their clothing."

On a less gross note, vitamin A, which is also necessary for the maintenance of the mucus-secreting epithelial cells that line (and protect) the respiratory tract and other vital organs, plays essential roles in maintaining the immune system and in preventing serious infections. It also prevents night blindness (and a condition called xerophthalmia, or nutritional blindness) and helps to form and maintain skin, teeth, and skeletal and soft tissues. Vitamin A breaks down easily when it's cooked or stored for a long time. And yes, it's true: high doses of beta-carotene, while not harmful, can temporarily give your skin a yellow or orange tint.

VITAMIN B1 (THIAMIN OR THIAMINE): *PROPOSED: 1906, ISOLATED: 1926, STRUCTURE DETERMINED: 1932, SYNTHESIS ACHIEVED: 1933*

Found in foods like yeast, enriched bread and flour products, eggs, lean and organ meats, legumes, nuts, seeds, peas, and whole grains, thiamin is water-soluble and sensitive to heat and alkaline conditions. It's also found in relatively high levels in the rinds of moldy cheeses like Brie and Camembert—approximately 0.4 mg/100 g, which is about ten times what's found in milk. Who said cheese wasn't good for you?

Thiamin is necessary for the enzymatic reactions that turn carbohydrates into energy, and plays important roles in the function of your heart, muscles, and nervous system. Severe thiamin deficiency causes a disease called beriberi that used to be widespread in countries where people's diets depended on white rice, since removing the rice's husk also removes its thiamin. Thanks in large part to the fortification and enrichment of flour and grain products, thiamin deficiency in the West is quite rare today. It's most often seen in severe alcoholics, in part because they tend to eat poorer diets to begin with, and in part because a lot of alcohol makes it hard for the body to absorb thiamin from food. Chronic thiamin deficiency in alcoholics often manifests itself as Wernicke-Korsakoff syndrome, which can cause confusion, memory loss, hallucination, a tendency to make up stories, and vision loss. It's also possible to inherit a genetic inability to absorb thiamin; this condition often develops over time, and since most Western doctors associate beriberi with alcoholism, it is often misdiagnosed. Thiamin's name refers to the fact that it contains sulphur—*thios* in Greek.

VITAMIN B2 (RIBOFLAVIN): *PROPOSED: 1933, ISOLATED: 1933, STRUCTURE DETERMINED: 1934, SYNTHESIS ACHIEVED, 1935*

Riboflavin plays essential roles in producing red blood cells and releasing energy from food, as well as maintaining the health of our skin and digestive tract. Riboflavin is naturally present in foods, including milk products, leafy green vegetables, and meat, and it's often added to breads and cereals via enriched flour. Interestingly, milk from cows fed on fresh grass has more riboflavin than that from cows fed on dry grass—which leads to natural seasonal fluctuations. Formerly known as vitamin G, riboflavin is heat-stable, so it doesn't break down if you cook it—but it does leach quite easily into cooking water, and degrades rapidly if it's exposed to light. Riboflavin deficiencies are extremely rare, but when they do occur their symptoms can include anemia, skin disorders, sores on your mouth or lips, a feeling of grittiness under your eyelids, and swelling of your mucous membranes. (And let us not forget, in extreme cases, vulval and scrotal dermatitis!) It would be extremely difficult to overdose on riboflavin, since it's not particularly well absorbed and is excreted in your urine. Speaking of which, if you have excess riboflavin, your urine will turn neon yellow (*flavus* is Latin for "yellow")—a harmless condition, which I find to be a useful indicator of whether I've taken a multivitamin.

VITAMIN B3 (NIACIN): *PROPOSED: 1926, ISOLATED: 1937, STRUCTURE DETERMINED: 1937, SYNTHESIS ACHIEVED: 1867*

Niacin is a water-soluble vitamin that's important for the digestive system, skin, and nerves; it also helps convert food to energy in cellular respiration. Niacin deficiency causes pellagra, the disease that was prevalent (in addition to other times and places) in the late 1800s and early 1900s in the American South, and which used to be called the disease of the four Ds: diarrhoea, dementia, dermatitis, and death. The highest natural sources of niacin are brewer's yeast and meat, but you can also find it in eggs, fish, legumes, nuts, and poultry, as well as in enriched breads and cereals. It's also in coffee beans, and it increases when they're roasted. Niacin was actually the first vitamin to be synthesized back in 1867, though no one knew it had a connection to nutrition—nicotinic acid, which is niacin's former name, was a common photographic chemical. (Bread producers insisted the name be changed to niacin so that customers didn't think their products had been enriched with nicotine.) It's very stable, and is not affected much by storage or cooking, but even in normal amounts, niacin can make your skin flush. Sometimes prescribed to lower cholesterol, it can also interact with many medications, especially blood thinners or medications for high blood pressure or diabetes.

VITAMIN B5 (PANTOTHENIC ACID): *PROPOSED: 1931, ISOLATED: 1939, STRUCTURE DETERMINED: 1939, SYNTHESIS ACHIEVED: 1940*

Pantothenic acid is a relatively stable water-soluble coenzyme that's important in the breakdown of fatty acids, amino acids, and carbohydrates, as well as for maintaining healthy skin. Heat-sensitive under certain circumstances, it's found in foods that are good sources of other B vitamins, especially organ meats, avocados, broccoli, mushrooms, and some yeasts; the highest sources, oddly, are in royal jelly (the special food of queen honeybees) and the ovaries of coldwater fish. No one yet understands how the body regulates its stores of pantothenic acid, but it seems like we may be able to somehow recycle it. It's extremely difficult for a non-starving person to be deficient in pantothenic acid—its Greek root, *pantothen*, means "from all sides," which gives a sense of how ubiquitous it is. However, it's worth noting that it's not present in common heavily refined foods like sugar, fats and oils, and cornstarch. D-panthenol, a provitamin of pantothenic acid (that is, a chemical our bodies can convert into the full-fledged vitamin), is a common cosmetic ingredient, since it is moisturizing and can make your hair shiny—in fact, its name was the inspiration for Pantene shampoo.

VITAMIN B6 (PYRIDOXINE): *PROPOSED: 1934, ISOLATED: 1936, STRUCTURE DETERMINED: 1938, SYNTHESIS ACHIEVED: 1939*

All forms of vitamin B6 are converted by our bodies into a coenzyme called pyridoxal phosphate, which plays a role in a surprisingly diverse array of functions and conditions including growth, cognitive development, depression, immune function, fatigue, and the activity of steroid hormones. It helps our bodies make antibodies and hemoglobin (the part of red blood cells that carries oxygen), maintains our nerves, and breaks down protein. Unless you're seriously malnourished, it's difficult to be deficient in B6, which is found in many of the same foods as the other B vitamins; the most concentrated sources include meats, whole-grain products (especially wheat), vegetables, and nuts. It can also be synthesized by bacteria, so moldy cheeses or cheese rinds can be good sources. In foods, B6 is stable if it's in acidic conditions, but otherwise it's sensitive to both heat and light.

VITAMIN B7 (BIOTIN): *PROPOSED: 1926, ISOLATED: 1939, STRUCTURE DETERMINED: 1942, SYNTHESIS ACHIEVED: 1943*

Along with pantothenic acid and pyridoxine, biotin is one of the B vitamins that no one really seems to care about. It's a water-soluble, fairly stable coenzyme that helps break down carbohydrates and fats and plays an important role in cellular respiration. But we need very little, and besides, biotin is present in so many foods, including brewer's yeast, eggs, nuts, sardines, whole grains, and legumes, that it's really hard to be deficient in it. People at the greatest risk of a biotin deficiency are pregnant women and people who have been fed by tube for a long time or who are generally malnourished. There isn't a good technique for measuring biotin in the body, so biotin deficiencies are usually identified by their symptoms, which include thinning hair and a scaly red rash around the eyes, nose, and mouth, as well as nervous system symptoms like depression, exhaustion, and hallucinations. Fun fact: Biotin was originally called vitamin H, after the German words *Haar* and *Haut*, meaning "hair" and "skin." Also: there's a substance in raw egg whites that binds to biotin in the intestine and keeps it from being absorbed by your body. If you'd like to give yourself a biotin deficiency, try eating two or more raw egg whites a day for several months.

VITAMIN B9 (FOLATE/FOLIC ACID): *PROPOSED: 1931, ISOLATED: 1939, STRUCTURE DETERMINED: 1943, SYNTHESIS ACHIEVED: 1946*

Folate, the synthetic form of which is known as folic acid, plays an essential role in closing an embryo's neural tube, the structure that will eventually become the baby's brain and spinal column. If a woman doesn't have enough of this vitamin at the time of conception, the tube may not completely close, leading to birth defects including spina bifida ("split spine"), which can cause nerve damage and paralysis to the legs, or anencephaly, a usually fatal condition whose name means "without a brain." The trick is that you need to make sure you're getting enough *before* you conceive, since the neural tube closes (and neural tube defects occur) before many women even know they're pregnant.

Today, all enriched grain products in the United States are required to include folic acid, in an attempt to prevent these birth defects. Since this became mandatory in 1998, the rate of neural tube defects in the United States has dropped by an estimated 25–50 percent. Nonetheless, more is not necessarily better (and thanks to mandatory enrichment, it can quickly add up)—too much folic acid can mask symptoms of a vitamin B12 deficiency.

Folate used to be called Wills' factor, after the woman who helped to discover it. Lucy Wills was a British doctor working with anemic pregnant women in Bombay—now Mumbai. Suspecting that their anemia might be related to poor nutrition, she experimented with a number of dietary changes and eventually discovered she could cure her patients with Marmite, the yeast-based spread—which we now know is rich in B vitamins—that's loved by the British and reviled by nearly everyone else. Folate helps the body make new cells, which is part of the reason why it's so important for pregnant women to have enough of it. Folate also works with vitamin B12 and vitamin C to help the body break down, use, and make new proteins, and also is essential in forming new red blood cells and replicating DNA.

Folate is prevalent in foods including leafy green vegetables (its name comes from *folium*, the Latin word for "leaf"), meats, dried beans, peas and nuts, citrus juices, and enriched breads and cereals, but it degrades if it's exposed to oxygen, and tends to leach into cooking water. In addition to birth defects, folate deficiency can cause diarrhoea, mouth ulcers, and certain types of anemia. The process of isolating it took four tons of spinach.

VITAMIN B12 (COBALAMIN): PROPOSED: 1926, ISOLATED: 1948, STRUCTURE DETERMINED: 1955, SYNTHESIS ACHIEVED: 1970

In many ways, vitamin B12 is the strangest of the vitamins. Deep red and crystalline in pure form, it is made almost exclusively by bacteria, including those that live in cows' rumens, sewage treatment plants, and the mud of the San Francisco Bay. It is the only molecule in the human body that contains cobalt (its alternate name is cobalamin) and is the best stored of the vitamins; even human babies, who are born with relatively low amounts of it, have enough to last for a year.

B12 also gives a sense of how recent the discovery of vitamins really is—indeed, some of the people who originally worked on isolating it are still alive. I had the chance to sit down with Gerald Combs Sr., an accomplished nutritionist who narrowly missed being credited for the 1948 isolation of vitamin B12 (and the father of Gerald Combs Jr., author of the textbook *The Vitamins*). "I was just a grad student doing a lot of hard work," he said when I asked him about his personal reaction to announcements from Merck and Glaxo Laboratories that they had isolated the vitamin. "But if I had crystallized it and been able to report it earlier, why, I would have been a famous little boy."

Relatively stable when cooked, each molecule of B12 consists of 181 atoms—the most of any vitamin, and a monster compared with vitamin C's 20 atoms. It took scientists twenty-three years to learn to synthesize it, compared with a mere three for folic acid. As a result of its complexity, synthetic vitamin B12 is made exclusively by microbial fermentation. Vitamin B12 is naturally found only in animal products like meat, fish, and dairy (the animals get the B12 from the microflora in their guts), which is why vegans are often deficient in it. Sources also include kidneys, liver, oysters, and—while I don't recommend this—faeces.

B12 is such a complicated molecule that it requires several steps to be absorbed. First, you need adequate gastric acid to cleave it from food. Then, once it has been separated, a special stomach secretion—called intrinsic factor—is required to make it available to your body. (B12 itself is known as extrinsic factor.) If you don't make enough gastric acid or can't make your own intrinsic factor, then you can't absorb B12 as it's found naturally in food and you may become deficient.

Indeed, if you're older than fifty, or a vegan, or if you take antacids or proton pump inhibitors, you may want to consider a supplement. The B12 from supplements is easier to absorb because it's not bound to food, and therefore doesn't require gastric acid or intrinsic factor in order to become available. If you're truly deficient, your doctor may recommend an injection of B12, which is the easiest form to absorb because it bypasses the stomach entirely. (This is only a good idea if you are actually deficient in B12; the recent fad of getting shots of B12 to "improve energy" is a waste of money.)

B12 is crucial for DNA synthesis, the maintenance of the central nervous system, and the formation of red blood cells. Symptoms of a B12 deficiency include everything from a loss of balance to hallucinations, disorientation, numbness or tingling in the hands and arms, memory loss, megaloblastic anemia, and irreversible dementia.[1] Thanks to its crucial role in the formation of blood cells, a B12 deficiency—whether it's caused by a lack of dietary sources or a genetic inability to make intrinsic factor—can also lead to a rare and formerly fatal condition called pernicious anemia (even more severe than its megaloblastic comrade).

The scientist who discovered intrinsic factor, William Bosworth Castle, MD, did so while trying to help patients who were dying of pernicious anemia. His method was both creative and one that would never be approved of today: he ate pieces of nearly raw meat, let them sit in his stomach till they were partially digested, regurgitated them, and gave the resulting slurry to his patients by tube. Thanks to the doctor's own intrinsic factor, the patients were able to absorb the B12; while they were not informed of its source, his vomit was their cure.

B12 can also be used as an antidote to cyanide poisoning.

VITAMIN C (ASCORBIC ACID): *PROPOSED: 1907, ISOLATED: 1926, STRUCTURE DETERMINED: 1932, SYNTHESIS ACHIEVED: 1933*

As you know by now, humans—along with guinea pigs, fruit bats, and some primates—are the only mammals that can't make their own vitamin C (in animals that can make their own, it's known as ascorbic acid). Vitamin C helps form collagen, a protein that your body uses to make skin, tendons, ligaments, and blood vessels; it also helps you heal wounds and form scar tissue and repair

---

[1] As a result, a B12 deficiency can have some odd symptoms. In one example of B12 deficiency: "A thirty-four-year-old woman developed a sore tongue and a sensation of electricity in her arms and legs. She thought that she smelled offensively, and told her children not to eat the food she had prepared. She became more depressed and tried to drown herself." In other words, you want to catch it early.

and maintain your cartilage, bones, and teeth—which is why scurvy can make your gums bleed and teeth fall out. It's also an important antioxidant.

Since it degrades quickly (and in response to practically anything), the best natural sources of vitamin C are fresh raw fruits and vegetables; it's particularly concentrated in citrus fruits like oranges and lemons, cantaloupe, kiwifruit, and various berries, broccoli, Brussels sprouts, cauliflower, bell peppers, leafy greens, and tomatoes. It's also abundant in sauerkraut.[2] Water-soluble, vitamin C is hard to overdose on, since you'll just excrete the extra in your urine—but most experts don't think that taking massive amounts of vitamin C will do anything for the common cold (let alone anything else). Also, people who smoke cigarettes tend to have lower levels of vitamin C.

Vitamin C is synthetically manufactured in higher quantities than any other vitamin, because it has so many other nonnutritional uses, including as a food additive to prevent unappetizing reactions like browning or off-flavours. But it also has industrial uses, including in photography, plastics manufacture, water treatment (to remove excess chlorine), stain removers, hair preparations, and skin treatments.

The story of vitamin C's discovery involves one of the most colourful characters in the history of vitamins: the Hungarian biochemist Albert Szent-Györgyi, who isolated vitamin C from oranges, lemons, cabbage, and adrenal glands (and later, paprika!) without knowing what it was. Szent-Györgyi was an extremely committed scientist. While serving as a medic in the Hungarian army during World War I, he shot himself in the arm so that he could go back to laboratory research.

My favourite detail about Szent-Györgyi is what he wanted to name the mysterious substance we now know as vitamin C. His first suggestion was "Ignose"—the "ose" for sugar, and the "ign" for his ignorance. When that was

---

[2] Captain James Cook intuited the value of sauerkraut as prevention for scurvy. From 1768 to 1771 he sailed the seas, taking astronomic measurements of the transit of Venus, trying to find the "southern continent" in the South Pacific, and forcing his men to eat sauerkraut along the way. In contrast to scurvy-stricken voyages like Anson's, Cook's didn't lose a single man to scurvy. The men didn't like sauerkraut, so Cook came up with the clever idea of having some specially prepared for the officers' table each day. "[T]he moment they see their superiors set a value on it, it becomes the finest stuff in the world," he observed.

rejected, he tried "Godnose" (say it out loud). When a humourless editor at the *Biochemical Journal* nixed that idea as well, he eventually named it hexuronic acid for its six carbons. Which, if you ask me, is really too bad.

VITAMIN D: *PROPOSED: 1919, ISOLATED: 1932, STRUCTURE DETERMINED: 1932 (D2), 1936 (D3), SYNTHESIS ACHIEVED: 1932 (D2), 1936 (D3)*

Unlike most vitamins, which help or participate directly in enzymatic reactions, vitamin D acts as a hormone—a chemical that tells the body to do something somewhere else. Also, unlike all the other vitamins, we don't need to get it from our diets, because our bodies make vitamin D in our skin with the help of UVB radiation from the sun. Indeed, it's likely that in the past, most of our vitamin D came from sun exposure, because most foods don't have very much of it; the best natural sources are fatty fish like tuna, salmon, and mackerel, and that old standby, cod-liver oil. (Vitamin D in milk is a synthetic addition.) Vitamin D is fat-soluble and relatively stable in foods, though it is sensitive to acids and—ironically, considering how our bodies naturally make it—light.

Our bodies need vitamin D in order to absorb calcium, a mineral that's crucial for normal bone formation—that's why vitamin D deficiency can cause rickets in kids, or soft bones (a condition known as osteomalacia) in adults.[3] (Vitamin D's role in regulating calcium absorption inspired one of its less healthy uses: as a rat poison.) Some scientists suspect that vitamin D may play a role in the prevention of other conditions, including cancer and type 1 diabetes, as well. But while these other potential effects are being actively researched, the vitamin D/calcium committee at the Institute of Medicine's Food and Nutrition Board concluded that at the time when they wrote their recommendations, bone health was the only condition with enough evidence behind it to use as an endpoint. As the committee itself indicates, this doesn't mean that calcium and vitamin D levels don't play a role in those other conditions; it just means that more work needs to be done.

---

[3] Speaking of vitamin D's effects on bone health, Michael Levine, MD, medical director of the Center for Bone Health at the Children's Hospital of Philadelphia, told me of a disturbing legal trend: defenders in severe child abuse cases using vitamin D blood tests to claim that rickets, not physical blows, had damaged the children's bones. Judges and juries take note: a low level of vitamin D in the blood does not prove that a person has vitamin D-related weakness in the bones; rickets is diagnosed not by blood but by X-rays examined by a trained radiologist.

While nothing is set in stone, many experts believe that if you live north of the line that connects Philadelphia, San Francisco, Athens, and Beijing, you probably don't get enough vitamin D from the sun, especially in the winter, and could likely benefit from a supplement. Likewise, if you don't spend much time outside (or if you always wear sunscreen, which blocks your body from making vitamin D), have dark skin, wear clothing that covers most of your body, or are older, overweight, or obese—all of which can negatively affect your body's production of and ability to use vitamin D—you may want to consider supplementation.

Also important to watch out for are vitamin D's interactions with certain prescription drugs—in particular, those that increase the production of the CYP3A4 enzyme in your liver. As mentioned, CYP3A4 is crucial for the proper metabolism of many medications, but it can also lower the levels of the active forms of vitamin D in circulation, which would mean you'd have to take more vitamin D to achieve the same levels in your blood. To find out if you're at risk of an interaction, Google your drug name and CYP3A4—and/or talk to your pharmacist.

There are two forms of vitamin D available in supplements—ergocalciferol (D2) and cholecalciferol (D3). D2 is made by plants; D3 is the form synthesized by animals (and by irradiating lanolin—it's the form in most supplements). According to Dr. Michael Levine, medical director of the Center for Bone Health at the Children's Hospital of Philadelphia, either form is fine if you're taking it daily. But if you're taking a once-a-week supplement—which is fine, since D is fat-soluble and thus won't wash out in your urine—he recommends taking D3, since the circulating form of vitamin D3 remains available in the blood longer than that of D2. As always, more is not always better: excessively high levels of D can make you absorb too much calcium, which can then end up in places it shouldn't, like your arteries. But don't worry about getting too much D from sunlight: our bodies seem to know when to stop producing it.

VITAMIN E: *PROPOSED: 1922, ISOLATED: 1936, STRUCTURE DETERMINED: 1938, SYNTHESIS ACHIEVED: 1938*

The term "vitamin E" is a catchall phrase for a family of substances—at least eight—that have the biological activity of alpha-tocopherol, a chemical substance whose name comes from the Greek words *tokos* ("childbirth") and *pherein* ("to bear"). It was discovered in part because of its role in preventing reproductive failure in rats. There are still many mysteries surrounding what it actually does for us, but we know that vitamin E is an extremely important fat-soluble antioxidant, able to protect cells from oxidative damage. (Vitamin E works along with water-soluble antioxidants like vitamin C to create a protective antioxidant network.) Its antioxidant qualities also make vitamin E a popular addition to foods and animal feeds to prevent spoilage. Alpha-tocopherol is the most bioavailable and bioactive form, and it, like all forms of vitamin E, is pale yellow and viscous in purified form. Vitamin E will darken if exposed to light, heat, or alkaline conditions, thanks to an oxidation reaction similar to the browning of sliced fruit. It becomes less stable at temperatures below freezing.

Vitamin E exists in some form in most photosynthetic organisms in the membrane of their cells, and tends to be concentrated more highly in the plant tissues that are exposed to light. Besides wheat germ oil, the richest natural sources are vegetable oils like corn, soybean, palm, sunflower, and safflower, and nuts and seeds. Despite the fact that it's fat-soluble, it's nearly impossible to overdose on vitamin E from food. And thanks to its natural abundance, it's also hard for most people to become deficient—so no need to worry about Elmer McCollum's alarming references to reabsorbed embryos.

VITAMIN K: PROPOSED: 1929, ISOLATED: 1939, STRUCTURE DETERMINED: 1939, SYNTHESIS ACHIEVED: 1940

The K comes from *koagulation*, the Danish word for "coagulation," which is an appropriate name, given this vitamin's important role in blood clotting. It's occasionally given by doctors to counteract medications that thin the blood—which means you shouldn't take it if you're actively trying to thin your blood with, say, Coumadin/warfarin. Vitamin K also plays a role in building strong bones. In nature, it's found in leafy green vegetables like kale, spinach, turnip greens, Swiss chard, parsley, and spring greens, as well as broccoli, cauliflower, cabbage, and Brussels sprouts. Small amounts of vitamin K also exist in fish, liver, meat, and eggs, and bacteria in your gastrointestinal tract can also make a little bit of the vitamin. Relatively heat-stable, vitamin K is fat-soluble, and outright deficiency is rare. It's often not included in multivitamin supplements.

CHOLINE

There is still disagreement over whether choline, which is present naturally in foods including eggs, beef liver, wheat germ, and cruciferous vegetables, should be considered a fourteenth human vitamin. (When it is, it's usually grouped with the B vitamins.) Here's what Gerald Combs, author of the textbook *The Vitamins*, had to say about it:

> Clearly, there are cases where animals who [are able to] make choline can still benefit from [additional] supplements of choline. By extension, I'd expect the same of some humans, namely, those with low intakes of protein, hence low intakes of methionine—the primary source of the labile methyl groups needed to make choline. I'd [also] expect choline to be of benefit to people with low food intakes due to sickness, inappetence, advanced aging, poverty, etc. Those groups are typically not specified in the RDAs, which are for the general population, [which is why there's no choline RDA] from the Institute of Medicine. . . . [But] choline is the only nutrient for which deprivation clearly enhances carcinogenesis. So, my personal take is that it is not smart for us to write it off.

Also worth noting: choline is a precursor to a chemical called trimethylamine. If you can't break down trimethylamine—which some people are genetically unable to do—your body odour may smell strongly of fish.

APPENDIX B:
ABBREVIATIONS AND DEFINITIONS

ABBREVIATIONS

AI	Adequate Intake
ARS	Agricultural Research Service
CRN	Council for Responsible Nutrition
DFE	Dietary Folate Equivalent
DRI	Dietary Reference Intake
EAR	Estimated Average Requirement
FDA	Food and Drug Administration
FNB	Food and Nutrition Board (at Institute of Medicine)
GMO	Genetically Modified Organism
GMPs	Good Manufacturing Practices

GRAS	Generally Recognized As Safe
IOM	Institute of Medicine
IU	International Unit
MDRIs	Military Dietary Reference Intakes
MRE	Meal, Ready to Eat
NAS	National Academy of Sciences
NCCAM	National Center for Complementary and Alternative Medicine
NHA	Nutritional Health Alliance
NHF	National Health Federation
NIH	National Institutes of Health
RAE	Retinol Activity Equivalent
RDA	Recommended Dietary Allowance
SNP	Single Nucleotide Polymorphism
UL	Tolerable Upper Intake Level
USDA	United States Department of Agriculture
WHO	World Health Organization

DEFINITIONS

Recommended Dietary Allowance (RDA): The average daily dietary nutrient intake level sufficient to meet the nutrient requirement of nearly all (97–98 percent) healthy individuals in a particular life stage and gender group.

Adequate Intake (AI): The recommended average daily intake level based on observed or experimentally determined approximations or estimates of nutrient intake by a group (or groups) of apparently healthy people that are assumed to be adequate—used when an RDA cannot be determined.

Tolerable Upper Intake Level (UL): The highest average daily nutrient intake level that is likely to pose no risk of adverse health effects to almost all individuals in the general population. As intake increases above the UL, the potential risk of adverse effects may increase.

Estimated Average Requirement (EAR): The average daily nutrient intake level estimated to meet the requirement of half the healthy individuals in a particular life stage and gender group.

(Committee on Use of Dietary Reference Intakes in Nutrition Labeling, *Dietary Reference Intakes: Guiding Principles for Nutrition Labeling and Fortification*)

Below are the current RDAs for vitamins from the Institute of Medicine at the National Academy of Sciences' Food and Nutrition Board, which is the non-governmental organization contracted by Congress to update the United States' nutritional recommendations.

Note, however, that the FDA, not the Food and Nutrition Board, is responsible for deciding which version of the RDAs should be used for the Nutrition and Supplement Facts panels that appear on food and supplement labels. Updates are pending, but as of late 2014, most of the "% DV" vitamin values on Nutrition and Supplement Facts panels are based on the 1968 version of the FNB's recommendations. This means that, until the pending updates are finalized and implemented, the "100%"s that you often see on Nutrition and Supplement Facts panels (under "% DV") are based on outdated recommendations. If you want to calculate what percentage of the up-to-date RDA a product contains, use this chart instead.

Dietary Reference Intakes (DRIs): Recommended Dietary Allowances and Adequate Intakes, Vitamins

Food and Nutrition Board, Institute of Medicine, National Academies

Life Stage Group	Vitamin A (µg/d)[a]	Vitamin C (mg/d)	Vitamin D (µg/d)[b,c]	Vitamin E (mg/d)[d]	Vitamin K (µg/d)	Thiamin (mg/d)	Riboflavin (mg/d)	Niacin (mg/d)[e]	Vitamin B$_6$ (mg/d)	Folate (µg/d)[f]	Vitamin B$_{12}$ (µg/d)	Pantothenic Acid (mg/d)	Biotin (µg/d)	Choline (mg/d)[g]
Males														
9–13 y	600	45	15	11	60*	0.9	0.9	12	1.0	300	1.8	4*	20*	375*
14–18 y	900	75	15	15	75*	1.2	1.3	16	1.3	400	2.4	5*	25*	550*
19–30 y	900	90	15	15	120*	1.2	1.3	16	1.3	400	2.4	5*	30*	550*
31–50 y	900	90	15	15	120*	1.2	1.3	16	1.3	400	2.4	5*	30*	550*
51–70 y	900	90	15	15	120*	1.2	1.3	16	1.7	400	2.4[h]	5*	30*	550*
> 70 y	900	90	20	15	120*	1.2	1.3	16	1.7	400	2.4[h]	5*	30*	550*
Females														
9–13 y	600	45	15	11	60*	0.9	0.9	12	1.0	300	1.8	4*	20*	375*
14–18 y	700	65	15	15	75*	1.0	1.0	14	1.2	400[i]	2.4	5*	25*	400*
19–30 y	700	75	15	15	90*	1.1	1.1	14	1.3	400[i]	2.4	5*	30*	425*
31–50 y	700	75	15	15	90*	1.1	1.1	14	1.3	400[i]	2.4	5*	30*	425*
51–70 y	700	75	15	15	90*	1.1	1.1	14	1.5	400	2.4[h]	5*	30*	425*
> 70 y	700	75	20	15	90*	1.1	1.1	14	1.5	400	2.4[h]	5*	30*	425*

[a] As retinol activity equivalents (RAEs). 1 RAE = 1 µg retinol, 12 µg β-carotene, 24 µg α-carotene, or 24 µg β-cryptoxanthin. The RAE for dietary provitamin A carotenoids is two-fold greater than retinol equivalents (RE), whereas the RAE for preformed vitamin A is the same as RE.

[b] As cholecalciferol. 1 µg cholecalciferol = 40 IU vitamin D.

[c] Under the assumption of minimal sunlight.

[d] As α-tocopherol. α-Tocopherol includes RRR-α-tocopherol, the only form of α-tocopherol that occurs naturally in foods, and the 2R-stereoisomeric forms of α-tocopherol (RRR-, RSR-, RRS-, and RSS-α-tocopherol) that occur in fortified foods and supplements. It does not include the 2S-stereoisomeric forms of α-tocopherol (SRR-, SSR-, SRS-, and SSS-α-tocopherol), also found in fortified foods and supplements.

[e] As niacin equivalents (NE). 1 mg of niacin = 60 mg of tryptophan; 0–6 months = preformed niacin (not NE).

[f] As dietary folate equivalents (DFE). 1 DFE = 1 µg food folate = 0.6 µg of folic acid from fortified food or as a supplement consumed with food = 0.5 µg of a supplement taken on an empty stomach.

g Although AIs have been set for choline, there are few data to assess whether a dietary supply of choline is needed at all stages of the life cycle, and it may be that the choline requirement can be met by endogenous synthesis at some of these stages.

h Because 10 to 30 percent of older people may malabsorb food-bound B_{12}, it is advisable for those older than 50 years to meet their RDA mainly by consuming foods fortified with B_{12} or a supplement containing B_{12}.

i In view of evidence linking folate intake with neural tube defects in the fetus, it is recommended that all women capable of becoming pregnant consume 400 μg from supplements or fortified foods in addition to intake of food folate from a varied diet.

j It is assumed that women will continue consuming 400 μg from supplements or fortified food until their pregnancy is confirmed and they enter prenatal care, which ordinarily occurs after the end of the periconceptional period—the critical time for formation of the neural tube.

NOTE: This table (taken from the DRI reports, see www.nap.edu) presents Recommended Dietary Allowances (RDAs) in **bold type** and Adequate Intakes (AIs) in ordinary type followed by an asterisk (*). An RDA is the average daily dietary intake level sufficient to meet the nutrient requirements of nearly all (97–98 percent) healthy individuals in a group. It is calculated from an Estimated Average Requirement (EAR). If sufficient scientific evidence is not available to establish an EAR, and thus calculate an RDA, an AI is usually developed. For healthy breastfed infants, an AI is the mean intake. The AI for other life stage and gender groups is believed to cover the needs of all healthy individuals in the groups, but lack of data or uncertainty in the data prevent being able to specify with confidence the percentage of individuals covered by this intake.

SOURCES: *Dietary Reference Intakes for Calcium, Phosphorous, Magnesium, Vitamin D, and Fluoride* (1997); *Dietary Reference intakes for Thiamin, Riboflavin, Niacin, Vitamin B_6, Folate, Vitamin B_{12}, Pantothenic Acid, Biotin, and Choline* (1998); *Dietary Reference Intakes for Vitamin C, Vitamin E, Selenium, and Carotenoids* (2000); *Dietary Reference Intakes for Vitamin A, Vitamin K, Arsenic, Boron, Chromium, Copper, Iodine, Iron, Manganese, Molybdenum, Nickel, Silicon, Vanadium, and Zinc* (2001); *Dietary Reference Intakes for Water, Potassium, Sodium, Chloride, and Sulfate* (2005); and *Dietary Reference Intakes for Calcium and Vitamin D* (2011). These reports may be accessed via www.nap.edu.

For the most up-to-date vitamin RDAs for other age groups (as well as pregnant and lactating women), visit:
http://www.iom.edu/Activities/Nutrition/SummaryDRIs/~/media/Files/Activity%20Files/Nutrition/DRIs/New%20Material/2_%20RDA%20and%20AI%20Values_Vitamin%20and%20Elements.pdf

For Tolerable Upper Intake Levels (i.e., the highest level of daily intake that hasn't been shown to cause adverse effects in most people), visit:
http://iom.edu/Activities/Nutrition/SummaryDRIs/~/media/Files/Activity%20Files/Nutrition/DRIs/ULs%20for%20Vitamins%20and%20Elements.pdf

NOTES

CHAPTER 1: High Seas and Hi-C

1 **"What is the function of these vitamines?"**: Harrow, Benjamin. *The Vitamines: Essential Food Factors*. New York: Dutton, 1922: 96.

1 **thirteen human vitamins:** As those lists indicate, vitamins are sometimes referred to by their chemical names—such as niacin—and sometimes by their letters, like vitamin K. Since the choice is often arbitrary, I'll use whichever name is the most familiar unless a distinction is necessary. Also, vitamins can come in different chemical forms, called vitamers, that have similar actions in the human body—for example, there's more than one form of vitamin D. Scientifically these distinctions can be important, since one vitamer can have more powerful effects or be more readily absorbed than another. But for our purposes, they're often more confusing than they're worth. Chemically speaking, vitamins are all organic compounds, which means they consist of more than one element and contain carbon, an element found in all living things; vitamin C, for example, is made of carbon, hydrogen, and oxygen. This makes them different from the dietary minerals, which are pure inorganic elements, meaning that they *don't* contain carbon—they're not produced by living things—and consist of just one element, like sodium or iron or potassium. Vitamins and minerals are often lumped together—both in our minds and in our multivitamins— but whereas dietary minerals are often incorporated into our bodies'

structures (your bones and teeth are primarily calcium, for example),
vitamins are not.

2 **less than the weight of two grains of Morton salt:** According to Morton Salt's
PR people, one grain is about 0.16 mg, or 160 mcg. According to its
nutritional label, a cube of Domino sugar is 2.5 g. Even the highest RDA—
which is for vitamin C—is just 90 mg per day, about one-thirtieth of the
weight of a Domino sugar cube. And C is extreme: the RDA of the second
highest, vitamin E, is 19 mg.

2 **Recommended Dietary Allowance for vitamin D:** Office of Dietary
Supplements' Vitamin D Health Professional Fact Sheet: http://ods.od.nih
.gov/factsheets/VitaminD-HealthProfessional/.

2 **B12, a vitamin whose deficiency:** Skerrett, Patrick. "Vitamin B12 Deficiency
Can Be Sneaky, Harmful." Harvard Health Blog RSS. http://www.health
.harvard.edu/blog/vitamin-b12-deficiency-can-be-sneaky-harmful
-201301105780.

3 **scurvy killed more than two million sailors:** Bown, Stephen. *Scurvy: How a
Surgeon, a Mariner, and a Gentleman Solved the Greatest Medical Mystery of the
Age of Sail.* New York: Thomas Dunne Books, 2003: 3.

3 **more deaths at sea . . . than all other diseases combined:** Ibid.: 5.

3 **"It rotted all my gums":** Ibid.: 34.

4 **"old biscuit reduced to powder":** Frankenburg, Frances Rachel. *Vitamin
Discoveries and Disasters: History, Science and Controversy.* Santa Barbara:
Praeger, 2009: 72.

4 **one of history's worst medical disasters:** http://www.britannica.com/
EBchecked/topic/26843/George-Anson-Baron-Anson.

4 **The idea that certain foods can cure:** Frankenburg, *Vitamin Discoveries and
Disasters:* 75.

4 **described certain vegetables:** Walter, Richard. *A Voyage Round the World.*
Edinburgh: Campbell Denovan, 1781: 130.

5 **our bodies . . . with the help of enzymes:** Enzymes are made, as are all
proteins, of long chains of amino acids—chemical molecules known as the
building blocks of protein—folded into three-dimensional shapes. Like keys
that fit specific locks, enzymes are highly specific in what they react with,
and this specificity is often indicated by their names: the suffix "ase" means
that a molecule is an enzyme, and the part before the "ase" gives a hint as to
what it does. Lactase, for example, is the enzyme that allows us to break
down lactose, the sugar in milk ("ose" endings typically refer to sugars).
While enzymes aren't affected or used up by the reactions they enable, they
can be denatured—unfolded and therefore inactivated—by heat or chemical
changes. Often this denaturing is permanent, which is one of the reasons our

bodies constantly need to create new enzymes (and, therefore, need a regular supply of vitamins).

6 "[today's nutritional recommendations] are not presented as such": Committee on Use of Dietary Reference Intakes in Nutrition Labeling, Food and Nutrition Board. *Dietary Reference Intakes: Guiding Principles for Nutrition Labeling and Fortification*. Washington, DC: The National Academies Press, 2003: 75.

6 "a continuum of benefits": Ibid.: 74.

6 the RDAs . . . are actually not meant to be personal: Every few years since 1943, the government has contracted the Food and Nutrition Board (which, as part of the Institute of Medicine, is an independent nonprofit with no congressional funding) to update the country's dietary recommendations to keep them in line with the latest nutritional science.

The Food and Nutrition Board's original goal—namely, creating recommendations that would prevent outright nutritional deficiencies— persisted for about thirty years. Then in the 1970s, evidence began to emerge suggesting that people's diets might play a role in reducing their risk of chronic conditions like cardiovascular disease and cancer. By the late 1980s, publications like the 1988 Surgeon General's Report on Nutrition and Health were describing significant links between diet and long-term health. As a result, the Food and Nutrition Board expanded its goal beyond the prevention of outright nutritional deficiencies to include using diet to prevent chronic health issues. In addition, the FNB began worrying that Americans' increasing use of dietary supplements, coupled with fortified products (and the 1976 Proxmire amendment, which made it illegal for the FDA to limit the amount of vitamins or minerals in dietary supplements), might lead to the risk of overconsumption of certain nutrients.

So in 1994, the Food and Nutrition Board successfully proposed that it issue its guidelines under a new structure—and abbreviation—called the Dietary Reference Intake, or DRI. Today's DRIs, which are a total of 5,000 pages long and which will continue to be updated based on new research, are issued in partnership with Canada and still include recommended dietary allowances, which are also still defined as the amounts of a given nutrient thought to be sufficient for nearly all (97–98 percent) of the healthy population, broken down into age and gender groups. But the DRIs also include three new estimates (and abbreviations): Tolerable Upper Intake Level (UL), Estimated Average Requirement (EAR), and Adequate Intake (AI). For their official definition, see appendix B.
(Committee on Use of Dietary Reference Intakes in Nutrition Labeling, *Dietary Reference Intakes: Guiding Principles for Nutrition Labeling and Fortification*.)

6 not established adult RDAs: Institute of Medicine. "Dietary Reference Intakes: Vitamins." http://www.iom.edu/Activities/Nutrition/

SummaryDRIs/~/media/Files/Activity%20Files/Nutrition/DRIs/5_
Summary%20Table%20Tables%201-4.pdf.

7 **Supposed triggers:** Frankenburg, *Vitamin Discoveries and Disasters*: 76.

7 **it's referred to as "ascorbic acid":** Before vitamins had been discovered or the idea of nutritional deficiency diseases had gained any sort of traction, two Norwegian scientists, Axel Holst and Theodore Frolich, randomly stumbled upon the experimental potential of the guinea pig when they were trying to study a disease known as ship beriberi. Unable to replicate the disease in pigeons, they switched to guinea pigs. After several weeks on an all-grain diet, the guinea pigs didn't have ship beriberi, but they'd developed the loose teeth and hemorrhages of scurvy—and Holst and Frolich had therefore discovered the first animal model for the condition. Holst and Frolich immediately switched their research focus to scurvy, and published a controversial paper in 1907, suggesting that it might be a nutritional deficiency disease. (Pure vitamin C was not isolated till 1926.)

8 **"most effectual remedies":** Lind, James. *A Treatise of the Scurvy: Containing an Inquiry into the Nature, Causes, and Cure, of That Disease. Together with a Critical and Chronological View of What Has Been Published on the Subject. 1st Edition.* Edinburgh: Kincaid & Donaldson, 1753: 196.

9 **"I do not mean to say that lemon juice":** Lind, James. *A Treatise on the Scurvy: Containing an Inquiry into the Nature, Causes, and Cure, of That Disease. Together with a Critical and Chronological View of What Has Been Published on the Subject. 3rd Edition.* London: G. Pearch and W. Woodfall, 1772: 526.

9 **Scurvy . . . even emerged among the babies:** Bown, *Scurvy*: 213.

10 **An estimated two billion people:** http://www.dsm.com/content/dam/dsm/cworld/en_US/documents/hidden-hunger.pdf.

10 **General vitamin deficiencies can and do occur:** Institute of Medicine. *Vitamin C Fortification of Food Aid Commodities: Final Report.* Washington, DC: The National Academies Press, 1997: 18. It's also worth noting the prevalence of devastating mineral deficiencies, including in iron and iodine.

CHAPTER 2: Plants and Plants

13 **"The discovery that tables may groan with food":** Levenstein, Harvey. *Paradox of Plenty: A Social History of Eating in Modern America.* Berkeley, CA: The University of California Press, 2003: 23.

13 **In 2011, the *Journal of Nutrition* published a report:** Fulgoni, Victor L., Debra R. Keast, Regan L. Bailey, and Johanna Dwyer. "Foods, Fortificants, and Supplements: Where Do Americans Get Their Nutrients?" *Journal of Nutrition* 141, no. 10 (2011): 1847.

14 **One of vitamins' roles in plants:** Among vitamins' roles in plants: phylloquinone (which we know as vitamin K1) helps transport energy (in

humans it helps blood clotting and bone formation), B vitamins like niacin, thiamin, and riboflavin are needed for enzymatic reactions (just like in humans), and folate (B9) is essential for regenerating the enzyme (ribulose bisphosphate) that fixes carbon dioxide from the air in the first stage of photosynthesis so that it can be transformed into energy, oxygen, and sugars.—Gene Lester, MS, PhD, research plant physiologist at the USDA Agricultural Research Service.

15 **our version contains a disabling mutation:** Zimmer, Carl. "Vitamins' Old, Old Edge." *New York Times*, December 10, 2013: D1.

15 **To get cod-liver oil:** Connor, John M. "The Vitamins Conspiracies: 1985–1999." In *Global Price Fixing*, Heidelberg: Springer (2007), revised April 9, 2008.

16 **most vitamins in supplements or enriched or fortified foods are synthetic, man-made substances:** Today, most vitamin fortification and enrichment is done with the help of vitamin premixes—blends of vitamins that food producers can buy or commission directly from vitamin manufacturers and then add to their products. That might sound simple, but creating perfect premixes is actually quite challenging: you must make sure that the vitamins in the mix (which can come in the form of a powder or a liquid) dissolve and/or blend properly, don't clump, don't add a bad taste, are present at least at the level claimed by the label, and make it through processing and storage intact, among many other technically tricky requirements. Liquids often require different premixes than dry foods, fat-soluble vitamins have different requirements than water-soluble, and certain foods lend themselves more naturally to being enhanced with particular vitamins than others (vitamins A and D go particularly well in milk, for example). Developing the perfect vitamin premix for each product is both an art and a science; that's why food manufacturers usually purchase premixes instead of trying to create their own.

16 **"It starts not with corn kernels or even corn starch":** Warner, Melanie. *Pandora's Lunchbox: How Processed Food Took Over the American Meal.* New York: Scribner, 2013: 84.

17 **"about as food-based as vitamins get":** Ibid.

17 **vitamin A's raw ingredients . . . niacin. . . . Thiamin:** Information on vitamin production is from a combination of Warner, *Pandora's Lunchbox*, and Connor, "The Vitamins Conspiracies": 31.

17 **lanolin, the greasy substance found in wool:** "Wool Grease Production: Going East." *Anhydrous Lanolin News*, Yixin Chemical Company. http://www.xinyilanolin.com/lanolinnews1.20.htm.

18 **"Deputy Secretary Wolfowitz Statement on Berets":** Wolfowitz, Paul. "Statement on Army Black Berets." http://www.defense.gov/news/May2001/d20010501beret.pdf.

18 **Berry Amendment was permanently enacted into law:** Grasso, Valerie Bailey. Congressional Research Service, "The Berry Amendment:

Requiring Defense Procurement to Come from Domestic Sources," July 20, 2012, 11. http://www.fas.org/sgp/crs/natsec/RL31236.pdf.

18 "[The] Berry Amendment . . . protection for critical industries": Ibid.

19 Roche still controlled . . . global vitamin market: Connor, "The Vitamins Conspiracies": 33.

19 These four companies produced nearly 80 percent: Ibid.: 19.

19 The scheme . . . had expanded: "Vitamin Cartel Fined for Price Fixing." *Guardian*, November 21, 2001.

19 "During the life of the conspiracy": Department of Justice. "F. Hoffmann-La Roche and BASF Agree to Pay Record Criminal Fines for Participating in International Vitamin Cartel." May 20, 1999. http://www .justice.gov/atr/public/press_releases/1999/2450.htm.

20 *Wall Street Journal* estimates: "Chinese Vitamin-C Suppliers Found Liable for Price-Fixing." *Wall Street Journal*, March 14, 2013.

20 "the most harshly punished antitrust violators": Connor, "The Global Vitamins Conspiracies": 8.

20 There hasn't been any American-led major producer: Ibid.: 17.

20 beta-carotene facility in Freeport, Texas: Confirmed via e-mail correspondence with Jean-Claude Tritsch at DSM, April 25, 2013.

20 vitamin E facility in Kankakee, Illinois: Confirmed via e-mail correspondence with BASF partner Caron Blitz, May 10, 2013.

20 doesn't manufacture any vitamins in America: E-mail correspondence with Jack Cox from Sanofi-Aventis, May 14, 2013.

20 Daiichi Fine Chemical: Leatherhead Food Research. *The Global Food Additives Market* (5th Edition). Surrey, UK: Leatherhead Food Research, 2011: 168; vitamin K information confirmed with Karen E. Todd, director of marketing, Kyowa Hakko USA, via e-mail correspondence, May 8, 2013.

20 Other Western producers . . . left the vitamin business: "Recent Shifts Among Vitamin Manufacturers Create New World Order of Fierce Competition." ICIS.com, February 28, 2005. http://www.icis.com/Articles/ 2005/02/25/654946/recent-shifts-among-vitamin-manufacturers-create-new- world-order-of-fierce.html. Today, there is only one large vitamin C plant operating outside China: DSM's facility in Dalry, Scotland. http://www .nutraingredients.com/Industry/DSM-makes-last-stand-against-Chinese- vitamin-C. The last vitamin C plant in America—a former Roche facility in Belvidere, NJ—closed in 2005, in part because of EPA complaints that it was emitting large quantities of toxic air pollutants including methanol and chloroform. ToxicRisk.com, "Roche Vitamins, Inc.," Belvidere, NJ. http:// toxicrisk.com/reports/15863/source.htm.

20 The global vitamin market: "Wool Grease Production: Going East," *Anhydrous Lanolin News.*

21 75 percent of vitamin D: D3 if you want to get technical. Other than vitamin C, these stats are from Michael Doyle, PhD, director of the Center for Food Safety at the University of Georgia, e-mail correspondence, April 21–22, 2013. Vitamin C stats are from Leatherhead Food Research, *The Global Food Additives Market* (5th Edition): 165.

21 70–90 percent: Ibid: 168.

21 Chinese vitamin C makers guilty of price-fixing: "U.S. Court Fines Chinese Vitamin C Makers." *New York Times*, March 15, 2013.

21 they'd be less likely to buy it: Israelsen, Loren. "Does 'Made in China' Matter to Supplement Consumers?" *Functional Ingredients*, July 18, 2011, http://newhope360.com/ingredients-general/does-made-china-matter-supplement-consumers.

22 it's not easy to open a new factory: Connor, "The Vitamins Conspiracies": 27.

22 get around the Berry Amendment . . . loophole: Exemptions also apply for certain materials that are not available in sufficient quantities from American manufacturers (goat hair canvas, metallic thread) or if they're for purposes deemed more important than patriotism ("acquisitions of waste and byproducts of cotton or wool fibre for use in the production of propellants and explosives," for example). Office of Textiles and Apparel (OTEXA), US Department of Commerce. "The Berry Amendment: Exceptions Provided Under the Berry Amendment When the Berry Amendment Does Not Apply": 2.

22 "We have the productive power": Darnton, Byron. "Experts Map Plan of Diet Education for Our Defense." *New York Times*, January 21, 1941.

23 "The government experts don't care": Ibid.

23 cornerstones of modern processed foods: "Major Crops Grown in the United States." *Environmental Protection Agency Ag 101:* http://www.epa.gov/agriculture/ag101/cropmajor.html.

23 more than four hundred new additives: Levenstein, *Paradox of Plenty:* 109.

23 "young brides whose talents are limited": Ibid.: 101.

23 the 1945 Chicken-of-Tomorrow contest: Ibid.: 109.

24 "If food isn't safe, convenient, good to eat": Ibid.: 113.

24 more than half of America's fresh fruit: USDA Economic Research Service, via Michael Doyle, PhD, director of the Center for Food Safety at the University of Georgia.

24 **"many Americans . . . recommended micronutrient"**: Fulgoni, Keast, Bailey, and Dwyer, "Foods, Fortificants, and Supplements: Where Do Americans Get Their Nutrients?": 1847.

25 **irradiating the milk with ultraviolet light**: Committee on Use of Dietary Reference Intakes in Nutrition Labeling, *Dietary Reference Intakes: Guiding Principles for Nutrition Labeling and Fortification*: 46.

25 **not getting their nutritional boost from fruit**: This reality came as a shock to a woman who brought a 2009 lawsuit against Cap'n Crunch claiming that the word "berries" misled consumers into thinking that the cereal contained real fruit. The judge, thankfully, did not buy her argument.

25 **it would be strange . . . *not* to fortify them**: In public health cases like this, the government encourages that certain foods be fortified or enriched based on several criteria: how naturally they lend themselves to the addition of that particular vitamin or nutrient (you want the new nutrient to be stable in the food to which it's being added, and in a form that's easily absorbable in people's bodies), whether the addition of the new nutrient will create an imbalance of other nutrients or a risk of overdose in the general population (the correct answer is no), and how frequently and widely the foods being fortified are consumed (you want your efforts to reach as many people as possible; that's why so many enrichment and fortification programs focus on staple foods). Committee on Use of Dietary Reference Intakes in Nutrition Labeling, *Dietary Reference Intakes: Guiding Principles for Nutrition Labeling and Fortification*: 47–48.

25 **modern grocery store as we know it**: Whereas in 1948 large grocery chains held only 35 percent of the country's grocery business, by 1963 it was almost half—an increase that would have been impossible without easy-to-transport foods with long shelf lives. America became so proud of its new grocery stores that in 1957, "when the U.S. government wanted to display 'the high standard of living achieved under the American economic system,' at the Zagreb Trade Fair, it reproduced a supermarket stocked with American processed foods and produce." Levenstein, *Paradox of Plenty*: 113.

25 **constant supply of synthetic vitamins**: As evidence of the influence of food manufacturers on the demand for synthetic vitamins, consider that by the end of the 1920s, the food processing industries were the largest of America's manufacturing industries, beating iron, steel, and textiles in terms of capital investment. Levenstein, *Revolution at the Table*: 151.

CHAPTER 3: Death by Deficiency

27 **"solicit[ed] that the part"**: Ridley, J. "An Account of an Endemic Disease of Ceylon, Entitled Berri Berri." In *The Dublin Hospital Reports and Communications in Medicine and Surgery, Volume the Second*, 227–253. Dublin: Hodges and McArthur, 1818: 228.

27 Beriberi's cause was a mystery: The word itself was equally mysterious. It
   could have come from the Sinhalese word *bhyaree*, meaning "weakness," or
   the Hindu *bharbari* ("swelling"). It could have been a combination of the
   Arabic terms *buhr* ("shortness of breath") and *bahri* ("marine"). Or perhaps it
   was from an Indonesian word that means "sheep," a reference to the victims'
   supposedly sheeplike gait. No one knows for sure. Carpenter, Kenneth J.
   Beriberi, White Rice, and Vitamin B: A Disease, a Cause, and a Cure. Berkeley:
   University of California Press, 2000: 25.

28 wild elephants: Ridley, "An Account of an Endemic Disease of Ceylon,
   Entitled Berri Berri": 239.

28 "it occurred, more than once": Ibid.: 234.

29 "I awoke with a sensation of tightness": Ibid.: 235.

29 forced to return to England: Carpenter, *Beriberi, White Rice, and Vitamin B*: 27.

29 "h[ad] not been completely restored": Ridley, "An Account of an Endemic
   Disease of Ceylon, Entitled Berri Berri": 236.

29 Scientific efforts to understand beriberi's cause: Carpenter, *Beriberi, White
   Rice, and Vitamin B*: 9.

30 Japanese naval officials ordered ships from Britain: Ibid.: 10.

30 roughly a third of enlisted men: Ibid.

31 an achievement for which he's still celebrated: Carpenter, Kenneth J. "A
   Short History of Nutritional Science: Part 2 (1885–1912)." *Journal of Nutrition*
   133 (2003): 978.

31 his dietary changes were not adopted by the army: Hawk, Alan. "The Great
   Disease Enemy, Kakke (Beriberi) and the Imperial Japanese Army." *Military
   Medicine* 171 (2006): 333–339. http://academia.edu/1837458/The_Great_
   Disease_Enemy_Kakke_Beriberi_and_the_Imperial_Japanese_Army.

32 besides a few fungal infections: Ackerknect, Erwin H. *A Short History of
   Medicine.* Baltimore, MD: Johns Hopkins University Press, 1982: 176.

33 chickens and pigeons are two of the only animals: Frankenburg, *Vitamin
   Discoveries and Disasters*: 18.

34 By November 1899: Carpenter, *Beriberi, White Rice, and Vitamin B*: 38.

34 "refused to allow military rice": Eijkman, Christiaan. "Nobel Lecture—
   Antineuritic Vitamin and Beriberi." Speech presented upon receipt of the
   1929 Nobel Prize in Physiology or Medicine, 1929. http://nobelprize/org/
   nobel_prizes/medicine/laureates/1929/eijkman-lecture.html.

35 "star-gazing": Combs, Gerald Jr. *The Vitamins: Fundamental Aspects in
   Nutrition and Health, 4th Edition.* London: Elsevier, Academic Press,
   2012: 272.

36 Thiamin deficiencies . . . common in alcoholics: Ibid.: 270.

36 Thiamin plays a crucial role: Medscape. "Beriberi (Thiamine Deficiency)." http://emedicine.medscape.com/article/116930-overview#a0104.

36 In its pure form . . . one-hundredth of a milligram: Carpenter, *Beriberi, White Rice, and Vitamin B*: 110.

36 his heart will be back to normal: Medscape. "Beriberi (Thiamine Deficiency)." http://emedicine.medscape.com/article/116930 -overview#aw2aab6b2b5aa.

36 half-life of between ten and twenty days: Combs, *The Vitamins*: 265–266.

37 beriberi bacterium in the white part of rice: Carpenter, "A Short History of Nutritional Science: Part 2 (1885–1912)": 979.

37 "anti-beriberi factor" in the rice polishings: Frankenburg, *Vitamin Discoveries and Disasters*: 19.

37 the rate went up to 1 out of 4: Carpenter, "A Short History of Nutritional Science: Part 2 (1885–1912)": 980.

38 "There occur in natural foods": Ibid.

38 statement didn't become widely known . . . twenty-five years: Carpenter, Kenneth. "The Nobel Prize and the Discovery of Vitamins." Nobelprize.org. http://www.nobelprize.org/nobel_prizes/medicine/articles/carpenter/index .html.

38 "[t]here is a still unknown substance in milk": Carpenter, *Beriberi, White Rice, and Vitamin B*: 102.

39 "[T]he colour of their skin changes suddenly to red": Frankenburg, *Vitamin Discoveries and Disasters*: 36. (In other words, you do not want pellagra.)

41 "Measured quantities of the materials": Goldberger, Joseph. "The Transmissibility of Pellagra: Experimental Attempts at Transmission to the Human Subject." *Public Health Reports* 31 (1916): 3166; Carpenter, Kenneth, ed. *Pellagra*. Stroudsburg, PA: Hutchinson Ross, 1981: 90–95.

41 seven cubic centimeters of blood: Kraut, Alan. *Goldberger's War: The Life and Work of a Public Crusader*. New York: Hill and Wang, 2003: 145.

41 "a pall of gloom over our fair Southland": Ibid.: 148.

41 ten thousand people in the United States died of it: Carpenter, *Pellagra*: xi.

42 "P-P" factor (short for pellagra-preventive factor): Kaempffert, Waldemar. "Authorities Sure of Pellagra Cure." *New York Times*, March 20, 1938.

42 "400,000 people succumb to pellagra in this country every year": Ibid.

43 **his boss's pet hypothesis:** Griminger, Paul. "Casimir Funk: A Biographical Sketch (1884–1967)." *The Journal of Nutrition* 102 (1972): 1108. http:// jn.nutrition.org/content/102/9/1105.full.pdf.

43 **a ton of rice bran:** Carpenter, *Beriberi, White Rice, and Vitamin B*: 110.

43 **Funk never completely isolated thiamin:** Rosenfeld, Louis. "Vitamine— Vitamin. The Early Years of Discovery." *Clinical Chemistry* 43:4 (1997): 681.

43 **called the mystery compound a "curative substance":** Funk, Casimir. "On the Chemical Nature of the Substance Which Cures Polyneuritis in Birds Induced by a Diet of Polished Rice." *Journal of Physiology* 43, no. 5 (1911): 395–400. http://www.ncbi.nlm.nih.gov/pmc/articles/PMC1512869/.

44 **"inclusion in one group, called** *deficiency diseases*": Funk, Casimir. "The Etiology of the Deficiency Diseases: Beriberi, Polyneuritis in Birds, Epidemic, Dropsy, Scurvy, Experimental Scurvy in Animals, Infantile Scurvy, Ship Beri-Beri, Pellagra." *Journal of State Medicine* 20 (1912): 341.

44 **"[T]he deficient substances . . . we will call 'vitamines'":** Ibid.: 342.

45 **"somewhat cumbrous nomenclature":** Rosenfeld, "Vitamine—Vitamin. The Early Years of Discovery": 681.

45 **"automatically fall into disuse":** McCollum, Elmer, and Cornelia Kennedy. "The Dietary Factors Operating in the Production of Polyneuritis." *Journal of Biological Chemistry* 24 (1916): 493.

45 **"soon join the 'musty company of phlogistic'":** Chittenden, Russell. "Story of the Vitamins." Russell Chittenden Papers, MS 611, Box 3, Folder 57, Manuscripts and Archives, Yale University Library.

46 **"the very term is pregnant with meaning":** Levenstein, Harvey. *Fear of Food: A History of Why We Worry About What We Eat.* Chicago: University of Chicago Press (2012): 80.

46 **"served as a catchword":** Carpenter, *Beriberi, White Rice, and Vitamin B*: 104.

CHAPTER 4: The Journey into Food

47 **"abundantly clear that before the last century closed":** Hopkins, Frederick Gowland. "Sir Frederick Hopkins—Nobel Lecture: The Earlier History of Vitamin Research." Nobelprize.org. Delivered December 11, 1929. http://www .nobelprize.org/nobel_prizes/medicine/laureates/1929/hopkins-lecture.html.

48 **should be written Calorie:** The word itself is derived from *calor*, Latin for "heat," which is one of the main by-products, along with water and carbon dioxide, that our bodies produce when converting food into energy. Technically speaking, a calorie with a lowercase "c" is the amount of energy necessary to raise a *gram* of water by one degree centigrade, and a Calorie with a capital "C," also known as a kcal, is the amount necessary to raise a

kilogram of water by one degree centigrade (from 14.5 to 15.5 degrees, if you really want to get specific). In America, we measure food in Calories/kcals, but no one other than nutritional scientists bothers to capitalize the word, which is why I've left it lowercase. Also, the energy in a calorie doesn't necessarily have to come from food; it could just as easily be from coal or wood or oil—indeed, the term "calorimetry" technically refers to the measurement of heat produced by *any* sort of chemical reaction.

48 **the ensuing rise in the water's temperature:** Painter, Jim. "How Do Food Manufacturers Calculate the Calorie Count of Packaged Foods?" *Scientific American*, July 31, 2006. http://www.scientificamerican.com/article.cfm?id=how-do-food-manufacturers.

49 **technically speaking, that's exactly what it means:** E-mail correspondence with Guy Crosby, science editor for America's Test Kitchen and adjunct associate professor at the Harvard School of Public Health, May 6, 2013.

49 **protein must be the most important nutrient:** Katch, Frank. "Sportscience History Makers—Liebig." SportSci.org, November 3, 1998. http://www.sportsci.org/news/history/liebig/liebig.html.

49 **That's why Eskimos ate so much fat:** University of Missouri. "Liebig's Dietetic Trinity: Food Revolutions: Science and Nutrition, 1700–1950." http://mulibraries.missouri.edu/specialcollections/exhibits/food/liebig.html; Brock, William H. *Justus von Liebig: The Chemical Gatekeeper.* Cambridge: Cambridge University Press, 2002: 184.

50 **body heat is produced regardless of which macronutrient:** Ibid.: 213.

50 **"My object . . . has been to direct attention":** Ibid.: 214.

50 **a Liebig-inspired cast:** These scientists included Max von Pettenkofer (1818–1901); Carl von Voit (1831–1908), a pupil of Liebig's; and Max Rubner (1854–1932).

52 **Atwater began evaluating:** Atwater, Wilbur O. "The Chemistry of Food and Nutrition." *Century*, May 1887: 62–64.

52 **"I must take a different view of food":** Atwater, Wilbur O. "What the Coming Man Will Eat." *Forum*, June 1892: 491.

52 **best diet . . . the cheapest and the most protein- and calorie-dense:** Levenstein, Harvey. *Revolution at the Table: The Transformation of the American Diet.* Berkeley: University of California Press, 2003: 45.

53 **"From the point of view of the peace of Atwater's soul":** Haller, Albert von. *The Vitamin Hunters.* Philadelphia, PA: Chilton, 1962: 36.

53 **"said by some that they never will be converted":** Lusk, Graham. *The Fundamental Basis of Nutrition.* New Haven: Yale University Press, 1914: 7.

55 **shared with . . . Christiaan Eijkman:** Hopkins shared the 1929 Nobel Prize in Physiology or Medicine with Christiaan Eijkman, the Dutch researcher

who established the connection between beriberi and polished rice (even though he was committed to finding a bacterial cause). It was a controversial choice, considering that neither Eijkman nor Hopkins ever actually isolated a vitamin.

56 **"abundantly clear that before the last century closed":** Hopkins, "Sir Frederick Hopkins—Nobel Lecture: The Earlier History of Vitamin Research."

57 **He began to get better within three days:** McCollum, Elmer. *From Kansas Farm Boy to Scientist.* Lawrence: University of Kansas Press, 1964: 14–16.

58 **chief chemist of the agricultural research station:** Ihde, Aaron. "Stephen Moulton Babcock—Benevolent Skeptic." In *Perspectives in the History of Science and Technology,* ed. Duane H. D. Roll. Norman: University of Oklahoma Press, 1971: 277.

58 **as a result of the 1887 Hatch Act:** Frankenburg, *Vitamin Discoveries and Disasters*: 1.

59 **Wisconsin lab's mission was to improve agriculture:** To its credit, the US Department of Agriculture had begun to recognize the importance of studying human nutrition. As it stated in 1896, "The time is not far distant when it will be generally recognized that man should pay at least as much attention to problems relating to his own food as to the study of food of domestic animals." "The Chemical Composition of American Food Materials," Bulletin No. 28, US Department of Agriculture, 1896.

59 **"something fundamental [about nutrition]":** McCollum, *From Kansas Farm Boy to Scientist*: 115.

59 **"They presented amazing contrasts":** Carpenter et al. "Experiments That Changed Nutritional Thinking." *Journal of Nutrition* 127, no. 5 (1997): 1017S–1053S.

60 **"It was a man-sized job":** Ibid.: 114.

60 **"pored over the journals of organic and biochemistry":** Ihde, "Stephen Moulton Babcock—Benevolent Skeptic": 278–279.

60 **at least thirteen papers . . . on the failure of purified diets:** McCollum, Elmer V. *A History of Nutrition: The Sequence of Ideas in Nutrition Investigations.* New York: Houghton Miflin, 1957: 201.

61 **"too wild, too much alarmed":** Ihde, "Stephen Moulton Babcock— Benevolent Skeptic": 280.

61 **Marguerite Davis:** Marguerite Davis had moved to Wisconsin to keep house for her widowed father. She was committed to helping her dad, but also didn't want to give up her own university studies; the head of the department of home economics had, for some reason, advised her to learn biochemistry. She showed up in McCollum's lab in 1909 to volunteer her services. For the next five years she worked full time for no salary (McCollum tried unsuccessfully

to get her one); in the sixth year, she finally received $600 for her help. She's credited as a coauthor on many of his early papers.

61 **something particular in the dairy fats:** You may notice that this is very similar to the observations of the British biochemist Frederick Hopkins. But McCollum and Davis came closer to isolating this substance than did Hopkins; they were also more assertive in claiming credit for the importance of their work.

61 **"certain accessory articles in certain food-stuffs":** McCollum, Elmer, and Marguerite Davis. "The Necessity of Certain Lipids During Growth." *Journal of Biological Chemistry* 15 (1913): 175.

62 **same compound that prevented and cured beriberi:** McCollum, Elmer, Nina Simmonds, and Walter Pitz. "The Relation of the Unidentified Dietary Factors, the Fat-Soluble A, and Water-Soluble B, of the Diet to the Growth-Promoting Properties of Milk." *Journal of Biological Chemistry* 27 (1916): 33.

62 **McCollum and his colleagues suspected:** McCollum and Kennedy, "The Dietary Factors Operating in the Production of Polyneuritis": 492.

62 **to alphabetical second place:** Carpenter, Kenneth J. "A Short History of Nutritional Science: Part 3 (1912–1944)." *Journal of Nutrition* 133 (2003): 3024.

62 **"unidentified dietary factor fat-soluble A":** Eddy, Walter, and Gessner Hawley. *We Need Vitamins: What Are They? What Do They Do?* New York: Reinhold, 1941: 9.

62 **discovered vitamins, period:** McCollum, *From Kansas Farm Boy to Scientist*: 134.

63 **seven Nobel Prizes:** Carpenter, Kenneth. "The Nobel Prize and the Discovery of Vitamins." Nobelprize.org, June 22, 2004. http://www .nobelprize.org/nobel_prizes/medicine/articles/carpenter/index.html

CHAPTER 5: From A to Zeitgeist

67 **Most of the vitamin A and D currently in milk:** Dairy Council. "Vitamins in Milk." http://www.milk.co.uk/page.aspx?intPageID=71.

68 *McCall's* **ran a feature in February 1935:** Splint, Sarah Field, and Camille Davied. "Are Your Menus Right?" *McCall's*, February 1935: 45, 68.

70 **"may be absorbed by mother's body":** McCollum, Elmer. "A Vitamin Primer." *McCall's*, April 1938: 60.

70 **deficiencies that did not . . . exist:** Levenstein, *Revolution at the Table*: 149.

70 **"The round shoulders":** McCollum, quoted by M. K. Wiseheart in an interview in *American Magazine*, January 1923: 14–15; 112.

71 **"In the vegetable kingdom":** Barnard, Eunice Fuller. "In Food, Also, a New Fashion Is Here." *New York Times Magazine*, May 4, 1930.

72 **"The food manufacturers have discovered a new language":** Eddy, Walter H. "How Do You Know You've Got a Vitamin?" *Good Housekeeping*, February 1929: 96.

72 **Sunkist lemons:** "For Refreshment, For Vitamin 'C.'" *McCall's* ad for Sunkist lemons, August 1930: 49.

72 **cod-liver oil:** "In Winter Months They Get So Little Sunshine." *McCall's* ad for Squibb's cod-liver oil, November 1928: 61.

72 **Iceberg lettuce:** "Take Internal Sun Baths Daily for Radiant Health." *McCall's* ad for iceberg lettuce, March 1930.

72 **bananas:** "This Natural Vitality Food." *McCall's* ad for bananas, April 1930: 51.

72 **Ralston Wheat Cereal:** *McCall's*, March 1941: 62.

72 **"Canned Pineapple":** *McCall's* ad for the Pineapple Producers Co-Operative Association, February 1922: 93.

72 **Del Monte . . . canned foods:** "Don't Risk Vitamin Starvation!" *McCall's* ad for Del Monte canned foods, February 1933.

72 **Vitamin D Beer:** "Schlitz with Vitamin D. . . . Whaaa?" *Wisconsinology*, January 25, 2009. http://wisconsinology.blogspot.com/2009/01/schlitz-with-vitamin-dwhaaa.html.

73 *S. cerevisiae* **feeds upon fermentable sugars:** Environmental Protection Agency. "*Saccharomyces cerevisiae* Final Risk Assessment." Biotechnology Program Under the Toxic Substances Control Act, February 1997. http://www.epa.gov/biotech_rule/pubs/fra/fra002.htm.

74 **up an additional 75 percent:** J. Walter Thompson Company. "Fleischmann Sales Continue Upward Trend," August 13, 1925. library.duke.edu/digitalcollections/mma_MM1148.

74 **"rich in *three* vitamins":** "Many Men Are Failures Because of *Intestinal Fatigue*." *McCall's*, September 1930: 135.

74 **"'XR' Yeast":** "This newly discovered Yeast is much *quicker-acting*." *McCall's*, December 1934: 119.

74 **"minerals and hormone-like substances":** "The New XR Yeast is biggest advance in treating constipation in a generation." *McCall's*, April 1935: 155.

74 **"[n]o other food, even fruits and vegetables":** "Completely corrected by this new discovery . . . chronic cases of constipation." *McCall's*, May 1935.

74 **Federal Trade Commission filed a cease-and-desist letter:** Levenstein, *Fear of Food*: 185.

270

75 **raise your skin's "self-disinfecting power":** These last few yeast claims are from "It's *remarkable how quickly* such Skin Eruptions disappear." *McCall's*, May 1932: 113.

75 **sharpen your intellect:** "Life Begins at 40." *McCall's*, December 1937: 119.

75 **prevent you from becoming fat:** Levenstein, *Revolution at the Table*: 198.

75 **restored a woman's ability to walk:** "Life Begins at 40." *McCall's*, November 1937: 131. Also, while the "Yeast for Health" campaign made yeast a popular food for human consumption, farmers were not convinced that it would do anything for their animals, as the Yeast for Health advertising agency found out in 1923 when it tried to market a dry form of yeast for poultry, swine, cattle, and other livestock. As the advertising agency's internal account reflects, the farmers "want[ed] positive proof of the efficacy of dry yeast feeding before they under[took] the additional trouble and expense it entails." They had "been made hostile to stock tonics and conditioners by a host of fake 'cures.'" Account History, The Fleischmann Company, by the J. Walter Thompson Company, January 2, 1926. http://library.duke.edu/digitalcollections/mma_MM1158/.

75 **would declare an unlimited national emergency:** *Proceedings: National Nutrition Conference for Defense, May 26, 27, 28, 1941, called by President Franklin D. Roosevelt.* Washington, DC: US Government Printing Office, 1942: xiii.

75 **"[t]o neglect food":** Ibid.: 230.

76 **"undaunted by mortality statistics":** Levenstein, Harvey. *Paradox of Plenty: A Social History of Eating in Modern America.* Berkeley: University of California Press, 2003: 59, 67.

76 **"[O]ur way of life will fail":** *Proceedings: National Nutrition Conference for Defense, May 26, 27, 28, 1941*: 1.

76 **the word "Food" should be swapped:** Darnton, "Experts Map Plan of Diet Education for Our Defense."

76 **"You cannot put into heavy industry":** Ibid.

76 **"extraordinary record in nailing Nazi aircraft":** *Proceedings: National Nutrition Conference for Defense, May 26, 27, 28, 1941*: 5.

77 **true reason for their improved accuracy:** World Carrot Museum. "Carrots in World War Two." http://www.carrotmuseum.co.uk/history4.html#nightvision.

77 **"unfortunate personality traits":** "Where does he get his disposition?" *McCall's* ad for Grape-Nuts, July 1927.

77 **"we can build a better and a stronger race":** *Proceedings: National Nutrition Conference for Defense, May 26, 27, 28, 1941*: 232.

77 **she believed came "very largely from the health":** Ibid.: 228.

78 "a systematic effort . . . German health and food authorities": "Hitler Vitamin Plan Is Nazi War Factor." *New York Times*, August 23, 1941.

78 "state of mental weakness and depression and despair": Levenstein, *Fear of Food*: 100.

78 "reach Mrs. Tom Jones in terms she can translate": Darnton, "Experts Map Plan of Diet Education for Our Defense."

78 nuanced, and hard-to-remember, distinctions: National Archives and Records Administration. "Preview | What's Cooking Uncle Sam?" http:// www.archives.gov/exhibits/whats-cooking/preview/kitchen.html.

79 "answering calls the rest of us don't hear": Altman, Alex. "America's Worst Vice Presidents." *Time*, 2008. http://www.time.com/time/specials/packages/ article/0,28804,1834600_1834604_1835417,00.html.

79 his "old friend, vitamin B1": *Proceedings: National Nutrition Conference for Defense, May 26, 27, 28, 1941*: 37.

80 "addition of the different types of vitamin B": Ibid.

80 In 1939, a crystallized form of B1: Apple, Rima. "Vitamins Win the War: Nutrition, Commerce, and Patriotism in the United States During the Second World War." In *Food, Science, Policy and Regulation in the Twentieth Century*, eds. David F. Smith and Jim Phillips. London: Routledge, 2000: 138.

80 *Journal of the American Medical Association* . . . report in 1940: "Vitamins for War." *Journal of the American Medical Association* 115, no. 14 (1940): 1198.

80 Wilder was no quack: Goldsmith, Grace. "Russell Wilder—A Biographical Sketch." *Journal of Nutrition* 74 (1961): 1–8.

81 greatest nutritional wartime threat: Russell Wilder and Elmer McCollum weren't the first people to be concerned about America's increasing taste for processed foods. In the 1830s, Reverend Sylvester Graham (of graham cracker fame) claimed that eating store-bought processed foods "contravened God's laws of health and contributed, among other things, to an epidemic of debilitating masturbation among the young." Levenstein, *Fear of Food*: 107.

81 supported the idea of fortifying fruits and vegetables: Levenstein, *Paradox of Plenty*: 68.

81 night blindness in Denmark in World War I: Semba, Richard. *The Vitamin A Story: Lifting the Shadow of Death*. Basel: Karger, 2012: 127.

81 thiamin as a "morale" vitamin: "News of Food: New Vitamin B-1 Chocolate Syrup Is Health Builder for Young and Old." *New York Times*, June 5, 1941.

81 "charm, composure and good digestion": "News of Food: Some Information on 'Enriched' Breads and Warnings on the Need for Vitamins." *New York Times*, January 6, 1942.

81 "stimulate without a letdown": "Food That Combats Fatigue." *New York Times*, October 18, 1942.

81 "vitalizing—and supposedly beautifying": "News of Food: Bread for Beauty Makes Appearance." *New York Times*, January 13, 1942.

81 "there is another most potent morale booster": Darnton, "Experts Map Plan of Diet Education for Our Defense."

82 "No Thiamin, no pep": "Give your family this needed protection." *McCall's* ad for Quaker Oats, October 1940.

82 "Wham" vitamin: Levenstein, *Fear of Food*: 101.

82 Vitamin Donuts: Levenstein, *Paradox of Plenty*: 75.

82 push millers to enrich their flour: Academy of Nutrition and Dietetics. "What's the Difference Between the Terms 'Fortified' and 'Enriched' on Food Labels?" EatRight.org. http://www.eatright.org/Public/content.aspx?id=6442453536#.URPiw1qjegQ.

82 by 1942, almost all America's bread: Levenstein, *Fear of Food*: 100.

82 most of America's flour is still enriched: While there is no mandatory flour fortification in the United States, there are requirements for what micronutrients must be added to flour and bread products that are labelled as "enriched"—a confusing use of the term, since contrary to the term's usual definition, enriched products often contain micronutrients at levels that the grain didn't originally contain. In the United States, enriched bread and flour must contain specific amounts of not just thiamin but iron, riboflavin, niacin, and since 1998, folic acid. (Calcium is optional.) Committee on Use of Dietary Reference Intakes in Nutrition Labeling. *Dietary Reference Intakes: Guiding Principles for Nutrition Labeling and Fortification*: 45. http://www.nap.edu/catalog.php?record_id=10872; "CFR—Code of Federal Regulations Title 21, Part 137: Cereal Flours and Related Products." Revised as of April 1, 2013. http://www.accessdata.fda.gov/scripts/cdrh/cfdocs/cfcfr/CFRSearch.cfm?fr=137.165.

83 "inconstant tenderness of the muscles of the calves": Levenstein, *Fear of Food*: 98.

83 "a feeling of unusual well-being": Williams, Ray, Harold Mason, Russell Wilder, and Benjamin Smith. "Observations on Induced Thiamin (Vitamin B1) Deficiency in Man." *Archives of Internal Medicine* 66, no. 4 (October, 1940): 785–799; Williams, Ray, Harold Mason, Benjamin Smith, and Russell Wilder. "Induced Thiamine (Vitamin B1) Deficiency and the Thiamine Requirement of Man: Further Observations." *Archives of Internal Medicine* 69, no. 5 (May 1942): 721–738.

83 "apathy was replaced by lively interest": Williams, Mason, Wilder, and Smith. "Observations on Induced Thiamin (Vitamin B1) Deficiency in

Man": 790–792; Williams, Mason, Smith, and Wilder, "Induced Thiamine (Vitamin B1) Deficiency and the Thiamine Requirement of Man."

83 "least desirable in a population facing invasion": "Vitamins for War." *Journal of the American Medical Association* 115, no. 14 (1940): 1198.

83 "more deficient . . . than commonly is reported in association with . . . beriberi": Williams, Mason, Wilder, and Smith. "Observations on Induced Thiamin (Vitamin B1) Deficiency in Man": 785; Williams, Mason, Smith, and Wilder. "Induced Thiamine (Vitamin B1) Deficiency and the Thiamine Requirement of Man": 721–738.

84 having them do chest presses: Levenstein, *Paradox of Plenty*: 21.

84 "in what was then called an insane asylum": Levenstein, *Fear of Food*: 99.

84 "vitamins are not 'just another food fad'": Darnton, "Experts Map Plan of Diet Education for Our Defense."

84 The hot vitamin of the early 1920s: Levenstein, *Paradox of Plenty*: 13. A deficiency in vitamin C supposedly led to tooth problems and put people at risk of becoming alcoholics; D was said to protect you from lead poisoning. And according to a researcher at the University of California, when rats were riboflavin-deprived, their hair turned grey; when they were given adequate amounts of the vitamin, it turned back to black—giving marketers fodder to tout riboflavin as a nutritional fountain of youth. Ibid., 14.

86 one of Roche's major sources of income: Connor, "The Global Vitamins Conspiracies": 16; Pfizer, Inc. "Exploring Our History: 1900–1950." http://www.pfizer.com/about/history/1900–1950. http://www.emmanuelcombe.org/great.pdf and http://www.amazon.com/Global-Fixing-Studies-Industrial-Organization/dp/3540786694.

86 the "vitamin gold rush of 1941": Ibid.

86 pep pills: Levenstein, *Paradox of Plenty*: 69.

86 "Vitamins for Victory": Apple, "Vitamins Win the War": 141.

86 employer-provided vitamins: Levenstein, *Fear of Food*: 102.

86 women equated vitamins with energy: Ibid.: 101.

87 "When the customer takes one of those": Robert W. Yoder in *Hygeia*, April 1942: 264–5 (reprinted from the *Chicago Daily News*).

87 vitamin sales . . . were about $136 million per year: Levenstein, *Fear of Food*: 102.

87 "a wave of nutritional reform": Kaempffert, Waldemar. "What We Know About Vitamins." *New York Times Magazine*, May 3, 1942: 10–11, 23.

87 "Get your vitamins in food": Levenstein, *Paradox of Plenty*: 20

88 a quarter of the average American's caloric intake: Ibid.: 22.

88 New measurement techniques . . . were revealing: Ibid.: 20.

88 co-opted Elmer McCollum: McCollum was not the only food scientist being co-opted by industry. Wilbur Atwater's daughter, Ruth Atwater—herself a trained nutritionist—was hired by the National Association of Canners to promote its products; another renowned researcher, Walter Eddy (who wrote his own column in *Good Housekeeping*), announced that the canned foods that he'd tested all contained more than sufficient amounts of A, B, C, and riboflavin, and that Jell-O was an excellent nutritional choice. Indeed, by the 1930s, the food industries had become mass-circulation magazines' biggest advertisers, hardly a recipe for journalistic objectivity. Levenstein, *Paradox of Plenty*: 15–16, 18.

88 paid by the National Bakers Association: Levenstein, *Fear of Food:* 88.

88 a $250,000 pledge (in 1938 dollars): Ibid.: 95.

89 "suicidal for the commercial enterprises involved": Levenstein, *Paradox of Plenty*: 21.

89 American Medical Association urged manufacturers: Ibid.: 21.

89 1933 ad for Cocomalt: "Food-drink gives new energy to thousands." *McCall's*, July 7, 1933: 69; "Tommy Needs Vitamins—and *I* Need an Adding Machine!" *McCall's*, March 3, 1940: 55. The 1933 advertisement notes in a parenthetical that the method for this addition is "under license by Wisconsin University Alumni Research Formation." That's because the process of creating vitamin D by irradiation had recently been patented by Harry Steenbock at the University of Wisconsin, via the Wisconsin Alumni Research Foundation (WARF), which he had created in hopes of raising money from university researchers' patents to fund further university research. WARF's most famous contribution to medicine is Coumadin, aka warfarin (get it?), an anticoagulant medication that University of Wisconsin scientists originally developed in the 1940s as a rodenticide—it makes rats bleed to death.

90 "at least equal, and often superior to, raw produce": Levenstein, *Paradox of Plenty*: 111.

CHAPTER 6: Nutritional Blindness

91 "[A]t least two billion": Klaus Kraemer, director of *Sight and Life*, quoted in http://globalfoodforthought.typepad.com/global-food-for-thought/2012/12/interview-with-dr-klaus-kraemer-on-tackling-malnutrition-and-micronutrient-deficiencies.html.

92 This final stage can take less than a day: Conversation with Alfred Sommer.

92 "The nocturnal blindness is at first partial": Semba, *The Vitamin A Story*: 3.

92 **silver nitrate to the tip of the penis:** This was the treatment suggestion of a German physician in 1841. Ibid.: 6.

93 **ancient Egyptians . . . night blindness:** Sommer, Alfred. "Nutritional Blindness: Xerophthalmia and Keratomalacia." *Duane's Ophthalmology*, vol. 5, ch. 59. http://www.oculist.net/downaton502/prof/ebook/duanes/pages/v5/v5c059.html.

93 **most well-nourished people have enough:** Semba, *The Vitamin A Story*: 69.

93 **one out of four preschool children:** West, Keith P. "Extent of Vitamin A Deficiency among Preschool Children and Women of Reproductive Age." Presented at the XX International Vitamin A Consultative Group (IVACG) Meeting, "25 Years of Progress in Controlling Vitamin A Deficiency: Looking to the Future," February 12–15, 2001, Hanoi, Vietnam. Published in *Journal of Nutrition* (2002): 2857S–2866S.

93 **other estimates are even higher:** World Health Organization. "Micronutrient Deficiencies: Vitamin A Deficiency." http://www.who.int/nutrition/topics/vad/en/.

93 **6.2 million pregnant women per year:** West, "Extent of Vitamin A Deficiency among Preschool Children and Women of Reproductive Age": 2857S–2866S.

94 **"It may be an inspiring thought":** Moore, Thomas. *Vitamin A*. Amsterdam: Elsevier, 1957: 263.

94 **vitamin's role in . . . mucosal epithelial linings:** Sommer, "Nutritional Blindness: Xerophthalmia and Keratomalacia."

96 **cost only about two cents:** Remarkably, the vitamin A itself only costs about half a cent—the rest goes to the gelatine capsule, which isn't even consumed since the oil is squirted into the kids' mouths. Sommer has been searching for cheaper packaging material for years.

98 **but not life-threatening:** Sommer recalls some health ministers saying that with yearly budgets of about $2 a child, they couldn't spare the 25 cents per child that it would take to buy and distribute twice-yearly vitamin A supplements—blindness was not a serious enough problem compared with everything else the children were facing.

98 **published them in the *Lancet* in 1983:** Sommer, Alfred, et al. "Increased Mortality in Children with Mild Vitamin A Deficiency." *Lancet* 322, issue 8530 (September 1983): 585–588. http://www.thelancet.com/journals/lancet/article/PIIS0140-6736(83)90677-3/abstract.

99 ***Lancet* published these findings in 1986:** Sommer, Alfred, et al. "Impact of Vitamin A Supplementation on Childhood Mortality. A Randomised Controlled Community Trial." *Lancet* 1, issue 8491 (May 1986): 1169–1173. http://www.ncbi.nlm.nih.gov/pubmed/2871418.

99 **a placebo control:** This is true, but the children in question were starting off extremely deficient—and obviously, given this deficiency, weren't at risk of

getting too much vitamin A through other sources. Also, the Indonesian government had actually forbidden Sommer from using a placebo control, since it was already in the midst of rolling out a nationwide vitamin A supplement program and considered the use of placebos to be immoral. Instead, Sommer compared villages that had begun the vitamin A program with those that had not. Woodward, Billy. *Scientists Greater Than Einstein: The Biggest Lifesavers of the 20th Century.* Fresno, CA: Quill Driver Books, 2009: 25.

99　"the chief function of vitamin A . . . is as anti-infective agent": Semba, *The Vitamin A Story*: 133.

100 58 percent: Ibid.: 142.

100 annual consumption of cod-liver oil: Ibid.: 147.

101 UNICEF estimates that these programs: "Goal: Reduce Child Mortality." UNICEF, http://www.unicef.org/mdg/childmortality.html.

101 one of the most cost-effective interventions: Frankenburg, *Vitamin Discoveries and Disasters*: 13.

101 number one investment the world could make: World Health Organization. "Press Release: New Paper on Ways to Fight Hunger and Malnutrition," April 9, 2008. http://www.copenhagenconsensus.com/sites/default/files/ Press_release_for_new_Bjorn_Lomborg_paper_on_hunger_and_ malnutrition.pdf.

101 up to 90 percent of those children may die: Alfred Sommer, e-mail correspondence, May 8, 2013.

102 if the beta-carotene occurs in a non-oily food: Conversation with Alfred Sommer, April 9, 2013.

102 The WHO tries to take this discrepancy: West, Clive E., Ans Eilander, and Machteld van Lieshout. "Consequences of Revised Estimates of Carotenoid Bioefficacy for Dietary Control of Vitamin A Deficiency in Developing Countries." *Journal of Nutrition* 132, no. 9S (September 2002): 2922S. http://jn.nutrition.org/content/132/9/2920S.full.pdf; conversation with Alfred Sommer, April 9, 2013.

103 slightly different conversion factors: Miller, Melissa, et al. "Why Do Children Become Vitamin A Deficient?" Proceedings of the XX International Vitamin A Consultative Group Meeting. *Journal of Nutrition* (2002): 2867S–2880S; West, Eilander, and van Lieshout. "Consequences of Revised Estimates of Carotenoid Bioefficacy for Dietary Control of Vitamin A Deficiency in Developing Countries": 2920S–2926S.

105 little black dots in their flesh: Fully formed banana seeds are the size of watermelon seeds; since they make bananas less enjoyable to eat, most bananas today are man-made, seedless clones, grown from cuttings. As a result, most bananas are genetically identical, which means that a single disease could potentially wipe out much of the world's supply. Indeed, this

fate befell the Gros Michel, aka Fat Michael or Big Mike, a popular banana variety that was virtually wiped out in the 1960s by the so-called Panama disease. The bananas we eat today are Gros Michel's replacement, a type called Cavendish bananas. Federoff, Nina. *Mendel in the Kitchen: A Scientist's View of Genetically Modified Foods*. Washington, DC: Joseph Henry Press, 2007: 17.

105 **a staple food for roughly half the world's population:** List from Ingo Potrykus, e-mail correspondence, May 7, 2013.

105 **many weaned babies eat little else:** Federoff, *Mendel in the Kitchen*: 2.

105 **growing up in post-World War II Germany:** Nash, J. Madeleine. "Grains of Hope." *Time*, February 12, 2001; e-mail, Ingo Potrykus, May 7, 2013.

106 **"the best that agricultural biochemistry has to offer":** Federoff, *Mendel in the Kitchen*: 7.

106 **"first compelling example of a genetically engineered crop":** Nash, "Grains of Hope."

107 **Fool's Gold:** Federoff, *Mendel in the Kitchen*: 7.

107 **There are many legitimate questions:** Talk by Margaret Mellon of the Union of Concerned Scientists, at the Knight Food Journalism Bootcamp at MIT, March 2013.

108 **as well as that of many other scientists:** Revkin, Andrew. "From Lynas to Pollan, Agreement That Golden Rice Trials Should Proceed." *New York Times*, Dot Earth, August 27, 2013. http://dotearth.blogs.nytimes.com/2013/08/27/from-mark-lynas-to-michael-pollan-agreement-that-golden-rice-trials-should-proceed/?_php=true&_type=blogs&_r=0.

108 **first non-greenhouse test field of golden rice:** Golden Rice Humanitarian Board. "The Golden Rice Project." www.goldenrice.org/Content1-Who/who2_history.php.

108 **five years of regulatory hurdles:** Ibid. One of the reasons it was not grown in a country that actually could benefit from golden rice is that part of the licensing agreement for using other biotech companies' patented processes required that the genetically modified crops be grown only in countries that have regulatory frameworks for them—which most developing countries did not.

108 **twenty-three times more beta-carotene:** Since there have only been two small tests of golden rice in human subjects, the question of how much of this beta-carotene can be converted into vitamin A in the human body is still open, but the results of a 2012 study run by a US-led research team in China looked promising. It found a remarkable 2:1 beta-carotene to vitamin A conversion ratio from the current variety of golden rice (GR2), which is as good as what can be achieved through straight oil-based beta-carotene, and

which would protect a child from severe deficiency with a normal-sized serving. But even if the conversion rate turns out to be lower (the only other human trial, of a small group of American adults, found a 3.8–1 conversion rate), the beta-carotene in the rice's endosperm, which itself is very easily digested, is still likely far more accessible than the equivalent amount of beta-carotene in leafy greens. According to the bioavailability study, 100–150 g of cooked golden rice (50 g dry) appeared to provide about 60 percent of the Chinese Recommended Nutrient Intake for vitamin A for six-to-eight-year-old children (the total recommendation is 700 mcg per day). This roughly correlates to about a cup of cooked golden rice. Tang, Guangwen, Yuming Hu, Shi-an Yin, Yin Wang, Gerard E. Dallal, Michael A. Grusak, and Robert M. Russell. "Beta-Carotene in Golden Rice Is As Good As Beta-Carotene in Oil at Providing Vitamin A to Children." *American Journal of Clinical Nutrition* 96 (2012): 658–664. http://www.goldenrice.org/ PDFs/GR_bioavailability_AJCN2012.pdf; Qiu, Jane. "China Sacks Officials Over Golden Rice Controversy." Nature.com, December 10, 2012. http:// www.nature.com/news/china-sacks-officials-over-golden-rice-controversy -1.11998.

109 **a complement to other strategies:** Golden rice has inspired other efforts to genetically engineer other vitamins into plants, like vitamins E, C, and folate. There are also other new beta-carotene-rich vegetables in development, including "orange cauliflower" and "yellow potato" (not to mention a "golden banana").

109 **Organizations like Greenpeace:** Greenpeace. "Golden Rice Illusion" website: http://www.greenpeace.org/international/en/campaigns/agriculture/ problem/genetic-engineering/Greenpeace-and-Golden-Rice/.

109 **vandalism of test fields:** Revkin, "From Lynas to Pollan, Agreement That Golden Rice Trials Should Proceed."

CHAPTER 7: From Pure Food to Pure Chaos

111 **"Not only must the consumer be not disfigured":** *Proceedings: National Nutrition Conference for Defense, May 26, 27, 28, 1941*: 8.

112 **including herbs and botanicals:** These definitions are from the Food and Drug Administration's web pages about dietary supplements—for example: http://www.fda.gov/Food/DietarySupplements/UsingDietarySupplements/ ucm109760.htm.

112 **more than eighteen thousand products:** Vitamin Shoppe's "About Us" page: http://www.vitaminshoppe.com/content/en/support/help/about_us.jsp.

115 **$32 billion in US sales:** "Highlights from the 2013 Supplement Business Report." *Nutrition Business Journal*, December 9, 2013. http://newhope 360.com/supplements/infographic-highlights-2013-supplement-business -report.

Notes

279

116 origins of America's supplement regulations: Hilts, Philip J. *Protecting America's Health: The FDA, Business and One Hundred Years of Regulation.* Chapel Hill: University of North Carolina Press, 2003: 17; e-mail correspondence with Suzanne Junod, PhD, FDA historian.

116 adulteration in American commerce: Hilts, *Protecting America's Health*: 22.

117 *Adulterations of Various Substances*: Beck, Lewis C. *Adulterations of Various Substances Used in Medicine and the Arts.* New York: S. S. and W. Wood, 1846.

117 same stigma that many Chinese products: Harris, Gardiner. "The Safety Gap." *New York Times*, October 31, 2008. http://www.nytimes.com/2008/11/02/magazine/02fda-t.html?_r=0.

117 European countries began embargoes: Hilts, *Protecting America's Health*: 30.

117 Bureau of Medicine and Surgery stopped buying drugs: Hilts, Philip J. "The FDA at Work: Cutting-Edge Science Protecting Consumer Health." *FDA Consumer*, Centennial Edition/January–February 2006.

117 snake oil: Hurley, Dan. *Natural Causes: Death, Lies, and Politics in America's Vitamin and Herbal Supplement Industry.* New York: Broadway Books, 2006: 23–25.

118 many of these "patent" medicines weren't patented: Hilts, *Protecting America's Health* (quoting James Harvey Young).

118 Wiley was the sixth of seven children: Hilts, *Protecting America's Health:* 13.

118 with piercing dark eyes and black hair: Young, James Harvey. "Two Hoosiers and the Two Food Laws of 1906." *Indiana Magazine of History* 88, no. 4 (December 1992): 309.

118 "Some said homely": Harding, Thomas Swann. *Two Blades of Grass: A History of Scientific Development in the US Department of Agriculture.* Norman: University of Oklahoma Press, 1947: 46.

118 enroll himself at Hanover College: Ibid.: 15.

118 "costume that included knee britches": Young, "Two Hoosiers and the Two Food Laws of 1906": 305–306.

118 "infamous bicycle incident of 1880": Hilts, *Protecting America's Health:* 16.

119 father of the beet sugar industry: Harding, *Two Blades of Grass*: 46.

119 "they often became unobtainable": Ibid.: 47.

119 Preservaline and . . . Freezem: Hilts. "The FDA at Work: Cutting-Edge Science Protecting Consumer Health."

119 no safety-testing requirements for these products: Ibid.

119 a dozen "young, robust fellows": Ibid.

120 borax-infused butter: List, Gary R. "Giants of the Past: Harvey W. Wiley."
    AOCS Lipid Library: http://lipidlibrary.aocs.org/history/Wiley/index.htm.
    Originally published in *Inform* 16, no. 2 (February 2005): 111–112.

120 After borax came salicylic acid: Hilts, *Protecting America's Health*: 40.

120 "Only the Brave Can Eat the Fare": Harding, *Two Blades of Grass*: 47.

120 *"If you ever visit the Smithsonian Institute"*: Janssen, Wallace F. "Food
    and Drug Administration Celebrates 75 Years of Consumer Protection:
    An Album from the Archives." *Public Health Reports* (1974–) 96, no. 6
    (Nov.–Dec. 1981): 490.

120 "an inability to perform work": Hilts, *Protecting America's Health*: 40.

121 symptoms that were similar or worse: Ibid.

121 a law that would keep chemical preservatives: Janssen, "Food and Drug
    Administration Celebrates 75 Years of Consumer Protection": 487–494.

121 The nascent food industry pushed back: Hilts, *Protecting America's
    Health*: 43.

121 infant "soothing syrups": Ibid.: 48.

121 American meat sales plummeted: Ibid.: 51; Young, James Harvey. "The Pig
    That Fell into the Privy: Upton Sinclair's *The Jungle* and the Meat Inspection
    Amendments of 1906." *Bulletin of the History of Medicine and Allied
    Sciences* 1985, 59: 467–480.

121 often called the Wiley Act: Young, "Two Hoosiers and the Two Food Laws
    of 1906": 318.

122 It required, among other things: Hilts, *Protecting America's Health*: 54.

122 1906 Pure Food and Drug Act had many loopholes: Ibid.

123 Congress had not authorized any money: Ibid.: 54.

123 powerful case in the *New York Times*: Campbell, Walter. "The New Food
    and Drug Bill." *New York Times*, July 9, 1933.

123 He wasn't exaggerating: Hilts, *Protecting America's Health*: 75.

123 less than the amount needed by the USDA: Ibid.: 71.

123 The exhibit, full of heartbreaking stories: Hilts. *Protecting America's
    Health*: 84.

124 Elixir Sulfanilamide contained 72 percent: Offit, Paul. *Do You Believe in
    Magic?: The Sense and Nonsense of Alternative Medicine*. New York:
    HarperCollins, 2013: 70.

125 had committed suicide: Hilts, *Protecting America's Health*: 92.

125 $26,000 (about $240 per death): Ibid.

125 **"My chemists and I deeply regret"**: Mihm, Stephen. "A Tragic Lesson." *Boston Globe*, August 26, 2007. Full quotation: http://www.fda.gov/About FDA/WhatWeDo/History/ProductRegulation/SulfanilamideDisaster /default.htm.

125 **called a "tonic" rather than an elixir**: "Death Drug's Hunt Covered Fifteen States." *New York Times*, November 26, 1937.

125 **Whereas in the 1920s**: Hilts, *Protecting America's Health*: 93.

126 **90 percent of prescriptions**: Ibid.: 105.

126 **drastically reduced the number of products**: Hilts, "The FDA at Work: Cutting-Edge Science Protecting Consumer Health."

127 **"FDA does not keep a list of manufacturers"**: Food and Drug Administration. "Q&A on Dietary Supplements 'Where can I get information about a specific dietary supplement?'" (Last updated March 20, 2014.) Though it is not reflected in this Q&A, manufacturers are now required to register with the FDA—but there is no way to tell if all have done so, and no master list exists of what supplements they produce. (Also, the list of registered manufacturers isn't public.) http://www.fda.gov/Food/DietarySupplements /QADietarySupplements/#where_info.

127 **which is required by the FDA**: Watson, Elaine. "Dan Fabricant: FDA 'Somewhat Aghast' at Degree of cGMP Non-Compliance." NutraIngredients-USA.com, April 26, 2012. http://www.nutraingredients-usa.com/Regulation/Dan-Fabricant-FDA-somewhat-aghast-at-degree-of -cGMP-non-compliance.

128 **Over-the-counter and pharmaceutical drugs**: Hilts, *Protecting America's Health:* 164.

129 **the most trusted in the world**: Food and Drug Administration. "The FDA's Drug Review Process: Ensuring Drugs Are Safe and Effective." http://www .fda.gov/Drugs/ResourcesForYou/Consumers/ucm143534.htm.

129 **As for the safety of America's *food* supply**: International Food Information Council. "Ensuring a Safe Food Supply: A Concise Guide to the US Food Regulatory System." http://ucfoodsafety.ucdavis.edu/files/ 26445.pdf.

130 **vitamins should be considered foods**: Levenstein, *Paradox of Plenty*: 20. According to the original version of the 1938 act, drugs were defined as substances listed in an official pharmacopeia; substances meant to diagnose, cure, mitigate, treat, or prevent disease; or substances (other than food) meant to affect the structure or any function of the body. The term "food," on the other hand, was defined as "(1) articles used for food or drink for man or other animals, (2) chewing gum, and (3) articles used for components of any such article" (vitamins appeared to fall under the third grouping). Why chewing gum warranted its own category is not immediately clear. Act of

June 25, 1937, Pub. L. No. 75–717, 52 Stat. 1040 ("Federal Food, Drug, and Cosmetic Act").

130 **In 1941, the FDA issued regulations:** Committee on the Nutrition Components of Food Labeling, Institute of Medicine. *Nutrition Labeling: Issues and Directions for the 1990s.* Washington, DC: National Academies Press, 1990: 57. http://www.nap.edu/openbook.php?record_id=1576&page=57.

130 **a 40 percent rise:** Hurley, *Natural Causes:* 41–45, 47.

131 **Vitamin A is by far the most dangerous:** The Tolerable Upper Intake Level for preformed vitamin A is 10,000 IU/day for healthy adults (3,000 mcg Retinol Activity Equivalents). Interestingly, high doses of vitamin A precursors like beta-carotene are *not* acutely toxic—the body seems to know when to stop converting them into vitamin A. Office of Dietary Supplements, Vitamin A Health Professional Fact Sheet: http://ods.od.nih.gov/factsheets/VitaminA-HealthProfessional/.

131 **early twentieth-century Antarctic explorer died:** Australian Government Department of the Environment. "Xavier Mertz. Home of the Blizzard—The Australasian Antarctic Expedition." http://mawsonshuts.antarctica.gov.au/cape-denison/the-people/xavier-mertz.

131 **high levels of vitamin E:** Office of Dietary Supplements Vitamin E Professional Fact Sheet: http://ods.od.nih.gov/factsheets/VitaminE-HealthProfessional/.

131 **routinely consuming too much vitamin D:** "FDA Letter to Industry Concerning Liquid Vitamin D Dietary Supplements." http://www.fda.gov/Food/GuidanceRegulation/GuidanceDocumentsRegulatoryInformation/DietarySupplements/ucm215527.htm.

131 **High levels of folic acid (B9) can mask signs:** Skerrett, "Vitamin B12 Deficiency Can Be Sneaky, Harmful."

131 **"function as free-floating drugs":** Herbert, Victor, ed. *The Mount Sinai School of Medicine Complete Book of Nutrition.* New York: St. Martin's Press, 1990: 90.

132 **The agency had issued proposed:** *Regulation of Dietary Supplements: Hearing Before the Subcommittee on Health and the Environment of the Committee on Energy and Commerce, House of Representatives on H.R. 509, H.R. 1709 and S. 784 Bills to Amend the Federal Food, Drug and Cosmetic Act to Establish Provisions and Standards Regarding the Composition and Labeling of Dietary Supplements.* 103rd Cong., 1st sess., July 29, 1993. Serial No. 103-157: 69-70 (Statements of David Kessler, Commissioner, Food and Drug Administration).

132 **FDA did not publish:** Nestle, Marion. *Food Politics: How the Food Industry Influences Nutrition and Health.* Berkeley: University of California Press, 2007: 237.

132 would be treated as an over-the-counter drug: Hurley, *Natural Causes*: 47.

133 roughly 25 percent of every consumer dollar: Aderholt, Robert. "Opening Statement, FY 2015 Budget Hearing, Food and Drug Administration." http://www.fda.gov/AboutFDA/WorkingatFDA/CareerDescriptions/ ucm112708.htm.

133 including food, pharmaceuticals, cosmetics: "About FDA: What Does FDA Do?" FDA website: http://www.fda.gov/AboutFDA/Transparency/Basics/ ucm194877.htm.

133 it must follow an official rule-making process: Office of the Federal Register. "A Guide to the Rule-Making Process." https://www .federalregister.gov/uploads/2011/01/the_rulemaking_process.pdf.

133 FDA did not have the authority: Nestle, *Food Politics*: 238.

134 two words: "No way": Ibid.

134 Vitamin-Mineral Amendment . . . Proxmire amendment: Nestle, *Food Politics*: 238.

134 "potentially harmful to their health": Hurley, *Natural Causes*: 48–49.

134 The FDA, whose position was backed: Ibid.: 53.

135 "what is overlooked by a great many people": Ibid.: 50.

135 "you cannot assume a product is safe": Ibid.: 51.

135 Marsha N. Cohen: Ibid.: 52.

136 Proxmire was ultimately more effective: Ibid.: 49.

136 "What the FDA wants to do": Ibid.

136 Proxmire amendment: Ibid.: 53. Text of Proxmire amendment (italics mine): "(A) the Secretary may not establish . . . maximum limits on the potency of any synthetic or natural vitamin or mineral within a food to which this section applies; (B) the Secretary may not classify any natural or synthetic vitamin or mineral (or combination thereof) as a drug solely because it exceeds the level of potency which the Secretary determines is nutritionally rational or useful; (C) the Secretary may not limit . . . the combination or number of any synthetic or natural—(i) vitamin, (ii) mineral, or (iii) *other ingredient of food*, within a food to which this section applies." Proxmire amendment: 21 U.S.C. § 350 ("Vitamins and Minerals"). http://www.gpo.gov/fdsys/pkg/USCODE-2010-title21/pdf/USCODE-2010-title21-chap9-subchapIV-sec350.pdf.

136 made it illegal for the FDA: Nestle, *Food Politics:* 238; Proxmire amendment: 21 U.S.C. § 350 ("Vitamins and Minerals").

137 "a charlatan's dream": Nestle, *Food Politics*: 238.

137 "the FDA . . . could have crippled us": Ibid.

137 *withdraw* **powers from the FDA:** Committee on the Nutrition Components of Food Labeling, Institute of Medicine. *Nutrition Labeling: Issues and Directions for the 1990s*: 57.

CHAPTER 8: The People's Pills

139 **"Recognize at the outset":** *Regulation of Dietary Supplements:* 63 (Statements of David Kessler, Commissioner, FDA).

139 **Supercritical Sea Buckthorn oil:** NewChapter.com. http://www.newchapter .com/targeted-herbal-formulas/supercritical-omega-7#supplement-facts.

140 **"After four months":** *Regulation of Dietary Supplements:* 139 (Statement of Dorothy C. Wilson).

141 **young reporter at the . . .** *Albuquerque Journal*: Hurley, *Natural Causes*: 59.

141 **a particular Japanese producer:** Slutsker, Laurence, et al. "Eosinophilia-Myalgia Syndrome Associated With Exposure to Tryptophan From a Single Manufacturer." *The Journal of the American Medical Association*, vol. 264, no. 2 (July 11, 1990). Also: Belongia, Edward et al. "An Investigation of the Cause of the Eosinophilia-Myalgia Syndrome Associated With Tryptophan Use." *The New England Journal of Medicine*, vol. 323, no. 6 (August 9, 1990).

141 **the exact cause of EMS:** Grady, Denise. "Dietary Supplement Found to Be Contaminated." *New York Times*, September 1, 1998. http://www .nytimes.com/1998/09/01/science/dietary-supplement-found-to-be -contaminated.html.

142 **"an accident waiting to happen":** *FDA's Regulation of the Dietary Supplement L-tryptophan: Hearing Before the Human Resources and Intergovernmental Relations Subcommittee of the Committee on Government Operations*. House of Representatives, 102nd Cong., 1st sess., July 18, 1991: 70 (Statement of Richard J. Wurtman, M.D., Massachusetts Institute of Technology, Cambridge, MA).

142 **"Tryptophan in dietary protein is an important nutrient":** Ibid.: 71 (Statement of Richard J. Wurtman, M.D., MIT).

142 **"Why didn't the FDA require a warning":** Hurley, *Natural Causes*: 66.

143 **"[M]y associates and I proposed":** *FDA's Regulation of the Dietary Supplement L-tryptophan*: 70 (Statement of Richard J. Wurtman, M.D., MIT).

144 **the FDA had withdrawn . . . regulations:** Nestle, *Food Politics*: 239; Health .gov. "Background on Dietary Supplements." http://www.health.gov/ dietsupp/ch2.htm#sthash.XmELnICr.dpuf.

144 **increased by 47 percent:** Nestle, *Food Politics*: 242.

144 **"cautious green light":** Ibid.

144 more than 40 percent of the new food products: Hilts, P. J. "In Reversal, White House Backs Curbs on Health Claims for Food." *New York Times*, February 9, 1990: A1, A22.

145 "not-for-supplement bias": Nestle, *Food Politics*: 242.

145 "significant scientific agreement": Ibid.: 250.

145 Silicon Valley of the supplement industry: Lipton, Eric. "Support Is Mutual for Senator and Utah Industry." *New York Times*, June 20, 2011.

145 "Without Senator Hatch": Daniells, Stephen. "Dietary Supplements Become Utah's #1 Industry, Topping $7.2 Billion." NutraIngredients-USA .com newsletter, May 23, 2012. http://www.nutraingredients-usa.com/ Markets/ietary-supplements-become-Utah-s-1-industry-topping-7.2-billion.

145 "he's our natural ally": Lipton, Eric. "Support Is Mutual for Senator and Utah Industry." *New York Times*, June 20, 2011.

145 sold vitamins as a young man: Ibid.

145 hundreds of thousands of dollars in political donations: Ibid.

145 left the task of actually *writing* those rules: Hurley, *Natural Causes*: 76.

146 with whom Kessler had worked before: Ibid.: 73.

146 no longer be what he called a "paper tiger": Ibid.

146 "It set off a firestorm": Phone interview with David Kessler, February 28, 2012.

146 founder and sole owner of Nature's Plus: Hurley, *Natural Causes*: 80.

146 passage of a pending bill . . . Waxman: Library of Congress Summary. "H.R. 2597 (102nd): Food, Drug, Cosmetic and Device Enforcement Amendments of 1991. Introduced June 7, 1991." https://www.govtrack.us/congress/bills/ 102/hr2597#summary/libraryofcongress.

147 "a war council unlike any before or since": Hurley, *Natural Causes*: 81.

147 NHA set a goal of raising $500,000: Ibid.: 83.

147 "if you make your voice be heard": Weisskopf, Michael. "In the Vitamin Wars, Industry Marshals an Army of Citizen Protesters." *Washington Post*, September 14, 1993: A07.

147 weekly letter to their congressmen: Hurley, *Natural Causes*: 87.

147 "Supports Freedom of Choice Regarding Natural Health Alternatives": Ibid.: 83.

147 "Write to Congress today": Nestle, *Food Politics*: 259.

148 "they just shrugged": Interview with Donna Porter, July 23, 2014.

148 "what the statute really means": Weisskopf, "In the Vitamin Wars, Industry Marshals an Army of Citizen Protesters": A07.

148 **2000 article in *HerbalGram***: Hurley, *Natural Causes*: 87.

148 **"lifeblood of the industry"**: Burros, Marian. "F.D.A. Is Again Proposing to Regulate Vitamins and Supplements." *New York Times*, June 15, 1993.

148 **"wacky even by Hill standards"**: Hilts, *Protecting America's Health*: 286.

149 **the agency obtained a warrant**: Hurley, *Natural Causes*: 85.

149 **posted a sign on its door**: Ibid.: 84.

149 **Given the owner's previous defiance**: Hilts, *Protecting America's Health*: 287.

149 **it was never pointed at anyone**: Ibid.

149 **that's not the story that went public**: Ibid.

150 **editorial in the *Seattle Post-Intelligencer***: Nestle, *Food Politics*: 255.

150 ***New York Times* . . . error-laden front-page article**: Williams, Lena. "F.D.A. Steps Up Effort to Control Vitamin Claims." *New York Times*, August 9, 1992.

150 **eighteen-paragraph correction on the *front page***: Williams, Lena. "A Correction: No Plan to Curb High-Potency Vitamins as Drugs." *New York Times*, August 16, 1992.

150 **two thousand letters were faxed to President Bush**: Nestle, *Food Politics*: 253.

150 **"For God's sake, we're talking about vitamin C"**: Williams, "F.D.A. Steps Up Effort to Control Vitamin Claims."

150 **Mel Gibson's medicine cabinet**: http://www.youtube.com/watch?v=IV2olDA0w8U; Hilts, *Protecting America's Health*: 288.

151 **Commissioner David Kessler telling the *New York Times***: Burros, "F.D.A. Is Again Proposing to Regulate Vitamins and Supplements."

151 **he didn't just want to block the FDA's proposed health-claim rules**: Hurley, *Natural Causes*: 89.

151 **Gerald Kessler . . . Dietary Supplement Health and Education Act**: Ibid.: 90–91.

152 **"When supplements are really drugs in disguise"**: *Regulation of Dietary Supplements:* 63 (Statements of David Kessler, Commissioner, FDA).

152 **"Think about it"**: Ibid: 63.

152 **Ninety-three percent of the salespeople had complied**: Ibid.

152 **Then he had assistants load the witness table**: "Diet Supplements Attacked by F.D.A." *New York Times*, July 30, 1993.

152 **"We are back at the turn of the century"**: *Regulation of Dietary Supplements:* 65 (Statements of David Kessler, Commissioner, FDA).

153 ***New York Times* . . . one far-fetched resolution**: Burros, "F.D.A. Is Again Proposing to Regulate Vitamins and Supplements."

153 **2002 Harris poll . . . more than half of Americans:** US Government
Accountability Office. *Dietary Supplements: FDA Should Take Further Actions
to Improve Oversight and Consumer Understanding*, GAO-09-250. Washington,
DC: US Government Accountability Office, 2009: 32.

154 **despite the fact that it was opposed:** Hurley, *Natural Causes*: 94.

154 **Dietary Supplement Health and Education Act:** Library of Congress
Summary of the 1994 Dietary Health and Education Act: https://www
.govtrack.us/congress/bills/103/s784#summary.

154 **Cosponsored by sixty-five senators:** Hurley, *Natural Causes*: 98-99.

154 **[it] was signed into law:** Full citation for DSHEA: Act of October 25, 1994,
Pub. L. No. 103–417, 108 Stat. 4325 ("Dietary Supplement Health and
Education Act of 1994"). Full text: http://www.gpo.gov/fdsys/pkg/
STATUTE-108/pdf/STATUTE-108-Pg4325.pdf.

154 **whose first draft had been penned:** Hurley, *Natural Causes*: 101.

154 **Clinton praised the legislation:** Health.gov summary of DSHEA: http://
www.health.gov/dietsupp/ch1.htm.

154 **Its first directive was to broaden:** "FDA Basics: 'What Is a Dietary
Supplement?'" http://www.fda.gov/aboutfda/transparency/basics/ucm195635
.htm.

154 **"hermaphroditic category":** Hurley, *Natural Causes*: 102.

154 **between food and drugs:** A further note on foods: At the time of this writing,
the FDA also hasn't established final guidelines for what qualifies an
ingredient as "generally recognized as safe" (GRAS). This is a problem
when old ingredients are used in formulations that, in the words of the
Government Accountability Office, have "drastically different safety profiles
than their historical use." Small amounts of bitter orange, for example, have
historically been used as a flavouring for orange marmalade—but bitter
orange also contains a powerful stimulant called synephrine that's related to
ephedrine, the active ingredient in the now-banned ephedra. Should a
company cite bitter orange's use in marmalade as justification to not submit
a new dietary ingredient notification to the FDA for a product that's 95
percent synephrine? Should it still be considered GRAS in this second,
potentially dangerous form? US Government Accountability Office, *Dietary
Supplements: FDA Should Take Further Actions to Improve Oversight and
Consumer Understanding*: 24.

155 **As journalist Philip Hilts writes:** Hilts, *Protecting America's Health*: 228.

155 **if a drug succeeds in the animal trials:** Food and Drug Administration.
"The FDA's Drug Review Process: Ensuring Drugs Are Safe and
Effective." http://www.fda.gov/Drugs/ResourcesForYou/Consumers/
ucm143534.htm.

155 **The FDA also inspects:** Food and Drug Administration. "Post Market
Surveillance Programs." http://www.fda.gov/Drugs/GuidanceCompliance
RegulatoryInformation/Surveillance/ucm090385.htm.

156 **more than ten thousand pages:** Hilts, *Protecting America's Health*: 229.

156 **the number of "new molecular entities" approved:** Food and Drug
Administration. "Is It True FDA Is Approving Fewer New Drugs Lately?"
http://www.fda.gov/downloads/AboutFDA/Transparency/Basics/
UCM247465.pdf.

156 **they're also time-consuming and worthy:** Berndt, Ernst. "PDUFA, Drug
Approval Times, Drug Safety Withdrawal Rates, and the Drug
Development Process, Empirical Findings." Presentation at the Prescription
Drug User Fee Act FDA Public Meeting at the National Institutes of
Health, November 14, 20015. http://www.fda.gov/ohrms/dockets/dockets/
05n0410/05n-0410-ts00006-Berndt.pdf

156 **Those that do succeed:** California Biomedical Research Foundation. "Fact
Sheet: New Drug Development Process." http://ca-biomed.org/pdf/media-
kit/fact-sheets/cbradrugdevelop.pdf; Lilly Investor FAQ: https://investor.lilly.
com/faq.cfm?faqid=3. The California Biomedical Research Foundation
estimates that the cost to bring a *single* drug from lab to market—not taking
the cost of failures into account—is about $359 million. Lilly, quoting
research from the Tufts Center for the Study of Drug Development, puts it
higher, at between $1.2 and $1.3 billion per drug.

156 **since for every drug that got approval:** *Forbes* found that companies that
focused exclusively on one drug spent a median of $311 million per drug, a
number that is artificially low, since it excludes the cost of failures (one-drug
companies whose drugs don't succeed often go out of business, so the $311
million figure only includes companies that happened to have everything go
right). *Forbes*'s calculations were based on non-inflation-adjusted financial
data from SEC filings (gathered by FactSet Systems) and approval data from
the FDA (gathered by the Innothink Center for Research in Biomedical
Innovation). The article's author, Matthew Herper, points out that some of
his numbers include the cost of post-market safety monitoring and R&D for
medical devices. But a 2012 estimate from the Tufts Center for the Study of
Drug Development, which attempted to exclude costs not directly related to
a drug's approval, still put the price of developing a new drug at $1.3 billion or
more. Herper, Matthew. "How Much Does Pharmaceutical Innovation Cost?
A Look at 100 Companies." *Forbes*, August 11, 2013; Herper, Matthew. "How
the Staggering Cost of Inventing New Drugs Is Shaping the Future of
Medicine." *Forbes*, August 11, 2013; Tufts Center for the Study of Drug
Development, January 26, 2012: http://csdd.tufts.edu/news/complete_story/rd
_pr_january_2012; direct correspondence with Matthew Herper, October 2013.

156 **DSHEA automatically grandfathered:** This might seem to make sense,
considering the potential costs and logistics of evaluating each of the 4,000

supplement products that were already on the market at the time of
DSHEA's passage. But such an enormous effort would have a precedent: in
1966, shortly after the thalidomide disaster, the FDA hired the National
Academy of Sciences to review the efficacy of the 4,000 or so drugs that had
been approved before the new tightened standards had gone into effect. The
effort, known as the Drug Efficacy Study, didn't require "strict standards of
evidence," but rather "some substantial evidence that each drug worked,"
according to journalist Philip Hilts. (Hilts, *Protecting America's Health*: 171.)
The study, which took three years to complete, found that 7 percent of
pharmaceuticals on the market were "completely ineffective" for every claim
they made, and that 50 percent were "effective to some degree on some
claims and ineffective on others." In the end, some 300 drugs were pulled
from the market—about 7.5 percent. Hilts, *Protecting America's Health*:
176–177; 4,000 supplement stat: US Government Accountability Office,
*Dietary Supplements: FDA Should Take Further Actions to Improve Oversight and
Consumer Understanding*: 1; http://www.fda.gov/Food/GuidanceCompliance
RegulatoryInformation/GuidanceDocuments/DietarySupplements/
ucm171383.htm#qa, accessed Oct. 21, 2011.

156 **requirements for *new* dietary ingredients:** Full text: "You are not limited in
what evidence you may rely on in determining whether the use of a new
dietary ingredient will reasonably be expected to be safe. [See section 413(a)
(2) of the act (21 U.S.C. 350b(a)(2)]. You must provide a history of use or other
evidence of safety establishing that the dietary ingredient, when used under
the conditions recommended or suggested in the labelling of the dietary
supplement, will reasonably be expected to be safe. To date, we have not
published guidance defining the specific information that the submission
must contain. Thus, you are responsible for determining what information
provides the basis for your conclusion. Nonetheless, we expect that—in
making a determination that a new dietary ingredient is reasonably expected
to be safe,—you will consider the evidence of safety found in the scientific
literature, including an examination of adverse effects associated with the use
of the substance." http://www.fda.gov/food/dietarysupplements/ucm109764
.htm. As mentioned in the text, the FDA did eventually publish draft guidance
on new dietary ingredients in 2011 defining the specific safety information
the FDA would look for in submissions, but at the time of this writing it had
not yet been finalized (and even final guidances are not binding).

156 **the long FDA approval process:** Herper, "How Much Does Pharmaceutical
Innovation Cost?" and "How the Staggering Cost of Inventing New Drugs
Is Shaping the Future of Medicine."

156 **DSHEA dictates that companies' only requirement:** Act of October 25, 1994,
Pub. L. No. 103-417, 108 Stat. 4325 ("Dietary Supplement Health and
Education Act of 1994"); http://www.fda.gov/RegulatoryInformation/
Legislation/FederalFoodDrugandCosmeticActFDCAct/Significant
AmendmentstotheFDCAct/ucm148003.htm.

156 **there's draft guidance pending:** In July 2011, the FDA issued draft
guidance (that is, proposed nonbinding recommendations) to clarify
what type of evidence it will look for to demonstrate the safety of new
dietary ingredients. By January 2012, the FDA had received more than
146,000 pages of comments, many of which were industry-generated—but
critics think that the new guidelines are still insufficient. For example,
the guidance would not require companies to submit unfavourable test
results, meaning that they could cherry-pick studies that suggested safety. It
also wouldn't require any studies in humans. Food and Drug Administration.
"New Guidance for Industry: Dietary Supplements: New Dietary
Ingredient Notifications and Related Issues. Nonbinding
Recommendations," July 2011. http://www.fda.gov/food/guidanceregulation/
guidancedocumentsregulatoryinformation/dietarysupplements/ucm257563
.htm. Concerns with FDA nonbinding recommendations on new dietary
ingredients: Cohen, Pieter. "Assessing Supplement Safety—The FDA's
Controversial Proposal." *New England Journal of Medicine* 366 (2012):
389–391. http://www.nejm.org/doi/full/10.1056/NEJMp1113325.

157 **requirements for new dietary ingredients . . . food additives:** "GRAS v. Food
Additive v. Dietary Ingredient." NaturalProductsInsider.com, August 4,
2011. http://www.naturalproductsinsider.com/news/2011/08/gras-v-food-
additive-v-dietary-ingredient.aspx.

157 **Food Safety Modernization Act:** Act of Jan. 4, 2011, Pub L. No. 111-353, 124
Stat. 3885 ("FDA Food Safety Modernization Act"). http://www.fda.gov/
Food/GuidanceRegulation/FSMA/.

157 **[FSMA] . . . granted the FDA the new authority to issue mandatory recalls:**
Fabricant, Daniel. "FDA Uses New Authorities to Get OxyElite Pro Off the
Market." *FDA Voice* blog, November 18, 2013. http://blogs.fda.gov/fdavoice/
index.php/2013/11/fda-uses-new-authorities-to-get-oxyelite-pro-off
-the-market/.

157 **FDA has banned only *one* dietary ingredient . . . ephedra:** 21 C.F.R.§119.1
("Dietary Supplements Containing Ephedrine Alkaloids"). http://www.gpo
.gov/fdsys/granule/CFR-2011-title21-vol2/CFR-2011-title21-vol2-sec119-1/
content-detail.html.

157 **deaths of more than a hundred people:** US Government Accountability
Office. "Testimony Before the Subcommittee on Oversight and
Investigations, Committee on Energy and Commerce, House of
Representatives: *Dietary Supplements Containing Ephedra, Health Risks and
FDA's Oversight.*" July 3, 2003. http://www.gao.gov/assets/120/110228.pdf.

159 **Even today you can still find:** 21 C.F.R.§119.1 ("Dietary Supplements
Containing Ephedrine Alkaloids").

158 **in the words of the Government Accountability Office:** US Government
Accountability Office. *Dietary Supplements: FDA Should Take Further Actions
to Improve Oversight and Consumer Understanding*: 26.

Notes

291

158 banning further dietary ingredients: http://www.fda.gov/Food/Recalls OutbreaksEmergencies/SafetyAlertsAdvisories/default.htm.

158 the preamble to DSHEA: Act of October 25, 1994, Pub. L. No. 103–417, 108 Stat. 4325 ("Dietary Supplement Health and Education Act of 1994"); http://www.fda.gov/RegulatoryInformation/Legislation/FederalFood Drugand CosmeticActFDCAct/SignificantAmendmentstotheFDCAct/ucm148003.htm.

158 deliberately restricted its official legislative history: Ibid. "Statement of Agreement: 'This statement comprises the entire legislative history for the Dietary Supplement Health and Education Act of 1994, S.784. It is the intent of the chief sponsors of the bill (Senators Hatch, Harkin, Kennedy, and Congressmen Richardson, Bliley, Moorhead, Gallegly, Dingell, Waxman) that no other reports or statements be considered as legislative history for the bill.'" (This statement is followed by five sentence-long clarifications.)

158 Smith Kline's . . . decision in the 1950s: Hilts, "The FDA at Work: Cutting-Edge Science Protecting Consumer Health": 60.

158 more than 85,000: Fabricant, Daniel. "FDA Uses New Authorities to Get OxyElite Pro Off the Market."

158 "The law still permitted": Hilts, *Protecting America's Health*: 61.

159 "likely be no dietary supplement industry": Council for Responsible Nutrition. "Dietary Supplements: Safe, Beneficial and Regulated." http://www.crnusa.org/CRNRegQandA.html; http://www.fda.gov/cosmetics/guidancecomplianceregulatoryinformation/ucm074201.htm.

160 do not have to be preapproved: Food and Drug Administration. "Structure/Function Claims." http://www.fda.gov/food/ingredientspackaginglabeling/labelingnutrition/ucm2006881.htm.

160 The compromise enabled: Hurley, *Natural Causes*: 101.

161 fundamentally absurd: A top FDA official (who did not wish to be identified) told a story illustrating this absurdity. She and a physician colleague were tasked with writing regulations for DSHEA, and they were struggling with trying to come up with an example of a legally allowable, but medically meaningless, claim. They finally arrived at "supports a healthy immune system." This was meant as a joke, but it has since become a common claim.

161 Structure/function statements: Food and Drug Administration. "Structure/Function Claims."
 Also, the disclaimer itself has an odd creation story. In order to get DSHEA passed before Congress recessed, Senator Tom Harkin (D-Iowa), a strong supporter of the supplement industry, asked Gerry Kessler if he was okay granting the FDA's demand for the bill to require the disclaimer to appear

on all supplement bottles. Kessler didn't want to make the decision himself, so he called several colleagues to get an outside okay on the disclaimer, eventually catching one on his car phone. (It was late on a Friday, so most people were out of the office.) That colleague was indecisive, but his wife piped up from the background to tell him to say it was okay. He did so, Kessler conveyed it to the lawmakers, and DSHEA was released from committee and passed. Hurley, *Natural Causes*: 101.

161 **few consumers know the difference:** France, Karen Russo, and Paula Fitzgerald Bone. "Policy Makers' Paradigms and Evidence from Consumer Interpretations of Dietary Supplement Labels." Presented at FDA Public Meeting: Assessing Consumers' Perceptions of Health Claims, November 17, 2005: slide 26. http://www.fda.gov/ohrms/dockets/dockets/05n0413/05n-0413-ts00007-France.pdf.

If a company wants to make a claim about its product that it worries will get it in trouble with the FDA, one option is to make the claim in an ad rather than on the supplement label—whereas labels are regulated by the FDA, ads are regulated by the Federal Trade Commission. Or it can lobby to have it included in one of the supplement almanacs frequently for sale near the vitamin aisle in drugstores, whose indexes often list herbs and supplements by the diseases they supposedly treat. Those are protected as free speech and aren't regulated by anyone.

161 **"We use focus groups":** Phone interview with David Kessler, February 28, 2012.

162 **Natural Curves:** GNC. "Biotech Corporation Natural Curves Breast Enhancement." http://www.gnc.com/product/index.jsp?productId=2134255.

163 **But *what* effects, exactly?:** http://nccam.nih.gov/health/chasteberry.

163 **supplement ingredients . . . health risks:** "Dangerous Supplements: What You Don't Know About These 12 Ingredients Could Hurt You." *Consumer Reports*, September 2010. http://www.consumerreports.org/health/natural-health/dietary-supplements/overview/index.htm. Dietary supplement ingredients banned by other countries: US Government Accountability Office. *Dietary Supplements: FDA Should Take Further Actions to Improve Oversight and Consumer Understanding*: 26.

164 **Comfrey, chaparral, germander, and kava:** National Center for Complementary and Alternative Medicine. "Time to Talk About Dietary Supplements: 5 Things Consumers Need to Know." http://nccam.nih.gov/health/tips/supplements?nav=gsa; *"Consumer Reports'* 'Dirty Dozen': 12 Risky Supplements." ABCNews.go.com, August 3, 2010. http://abcnews.go.com/Health/AlternativeMedicineSupplements/consumer-reports-dirty-dozen-12-risky-supplements/story?id=11309450.

164 **Kava has been banned:** "Dangerous Supplements Still at Large." *Consumer Reports*, May 2004: 12–17. http://consumersunion.org/pub/0504%20DietarySup.pdf.

164 **Aloe vera extract:** National Center for Complementary and Alternative Medicine. "Herbs at a Glance: Aloe Vera," updated April 2012. http://nccam.nih.gov/health/aloevera.

164 **"Herbs at a Glance":** National Center for Complementary and Alternative Medicine, "Herbs at a Glance." http://nccam.nih.gov/health/herbsataglance.htm.

164 **more than 50 percent of American adults reported:** Gahche, Jaime, Regan Bailey, Vicki Burt, Jeffery Hughes, Elizabeth Yetley, Johanna Dwyer, Mary Frances Picciano, Margaret McDowell, and Christopher Sempos. "Dietary Supplement Use Among U.S. Adults Has Increased Since NHANES III (1988–1994)." Centers for Disease Control and Prevention, NCHS Data Brief, April, 2011. http://www.cdc.gov/nchs/data/databriefs/db61.htm.

165 **St. John's wort could interact with your prescription drugs:** Center for Science in the Public Interest. "Warning Label Urged for St. John's Wort," November 10, 2011. http://www.cspinet.org/new/201111101.html.

165 **he compared St. John's wort to grapefruit:** http://www.medicinenet.com/script/main/art.asp?articlekey=14760.

166 **Of 335 residents and attending physicians:** Ashar, B. H., T. N. Rice, and S. D. Sisson. "Physicians' Understanding of the Regulation of Dietary Supplements." *Archives of Internal Medicine* 167 (May 2007): 966–969. http://www.ncbi.nlm.nih.gov/pubmed/17502539.

166 **many people conceal their supplement use:** American Society for Therapeutic Radiology and Oncology. "Cancer Patients Hide Their Use of Complementary and Alternative Treatments from Their Doctors." *Science Daily*, October 17, 2005. http://www.sciencedaily.com/releases/2005/10/051017072528.htm.

167 **It's also common . . . to substitute herbs:** Callaway, Ewen. "Chinese Medicine Herbs Found to Contain Endangered Species." ScientificAmerican.com, April 12, 2012. http://www.scientificamerican.com/article.cfm?id=chinese-medicine-herbs-found-to-contain-endangered-animals.

167 **the term *jin qian cao*:** "Chinese Herbs: Some Things to Remember." Veterinary.com, December 2, 1999. http://www.veterinarywatch.com/Chiherb8.htm.

167 **In 2013, Canadian researchers:** O'Connor, Anahad. "Herbal Supplements Are Often Not What They Seem." *New York Times*, November 3, 2013.

167 **DNA analysis of supposedly herbal ingredients:** Callaway, "Chinese Medicine Herbs Found to Contain Endangered Species"; for a list of other substances: "Chinese Herbs: Some Things to Remember."

167 **"such adventurous mixtures of multiple ingredients":** Callaway, "Chinese Medicine Herbs Found to Contain Endangered Species."

168 The Canadian research mentioned above: O'Connor, "Herbal Supplements
    Are Often Not What They Seem."

168 a piece of a Viagra tablet tumble out: Phone conversation with James
    Neal-Kababick, January 25, 2013.

168 "Supplement 411": US Anti-Doping Agency. "Supplement 411" web page:
    http://www.usada.org/supplement411.

168 an open letter to the supplement industry: Hamburg, Margaret. "Dear
    Manufacturer of Dietary Supplements" letter to supplement manufacturers
    about adulterants. Department of Health and Human Services, December
    10, 2010. http://www.fda.gov/downloads/Drugs/ResourcesForYou/Consumers/
    BuyingUsingMedicineSafely/MedicationHealthFraud/UCM236985.pdf.

168 an alarming variety of undeclared active ingredients: http://www.fda.gov/
    downloads/Drugs/ResourcesForYou/Consumers/BuyingUsingMedicine
    Safely/MedicationHealthFraud/UCM236985.pdf; anabolic steroids in
    vitamins: https://www.consumerlab.com/recall_detail.asp?recallid=10515.

169 FDA doesn't inspect contract labs: Watson, Elaine. "Dan Fabricant: FDA
    'Somewhat Aghast' at Degree of cGMP Non-Compliance."
    NutraIngredients-USA.com, April 26, 2012. http://www.nutraingredients-
    usa.com/Regulation/Dan-Fabricant-FDA-somewhat-aghast-at-degree-of-
    cGMP-non-compliance. James Neal-Kababick at Flora Labs also told me
    about this problem.

169 contaminants like arsenic, lead, and pesticides: Multiple sources, including
    interview with ConsumerLab president Tod Cooperman on January 25,
    2012. You can also search ConsumerLab.com for those terms (and others)
    and pull up relevant reports.

170 overages high enough to be a concern: Ibid.

170 more than two hundred people were poisoned: Offit, *Do You Believe in
    Magic?*: 91.

170 Bioterrorism Act of 2002: Act of June 12, 2002, Pub. L. No. 107-88, 116
    Stat. 594 ("Public Health Security and Bioterrorism Preparedness and
    Response Act of 2002") at § 305 ("Registration of Food Facilities").
    http://www.gpo.gov/fdsys/pkg/PLAW-107publ188/pdf/PLAW-107
    publ188.pdf.

170 Food Safety Modernization Act: Act of Jan. 4, 2011, Pub L. No. 111-353,
    124 Stat. 3885 ("FDA Food Safety Modernization Act"). http://www.fda.gov/
    Food/GuidanceRegulation/FSMA/.

170 FDA . . . knows the names and contact information: Food and Drug
    Administration. "Protecting the U.S. Food Supply: What You Need to
    Know About Registration of Food Facilities." The Public Health
    Security and Bioterrorism Preparedness and Response Act of 2002,

November 2003. http://www.fda.gov/downloads/Food/GuidanceRegulation/
UCM113877.pdf.

170 **aforementioned power of mandatory recall:** Daniel, "FDA Uses New
Authorities to Get OxyElite Pro Off the Market."

170 **2006 Dietary Supplement and Nonprescription Drug Consumer Protection
Act:** Act of December 22, 2006, Pub. L. 109–462, 120 Stat. 3469 ("Dietary
Supplement and Nonprescription Drug Consumer Protection Act").

170 **GMPs for supplements:** "Current Good Manufacturing Practice in
Manufacturing, Packaging, Labeling, or Holding Operations for Dietary
Supplements"; Final Rule, 72 Fed. Reg. 34751 (June 25, 2007).

171 **"serious adverse event":** http://www.gpo.gov/fdsys/pkg/PLAW-109publ462/
html/PLAW-109publ462.htm

171 **These "nonserious adverse event" reports:** "Dangerous Supplements: Still at
Large," *Consumer Reports*; Hilts, *Protecting America's Health*: 89; http://www.
gpo.gov/fdsys/pkg/FR-2007-10-15/html/07-5074.htm.

171 **no way for the FDA to identify and contact . . . suppliers or sellers:** Interview
with Daniel Fabricant, February 17, 2012; Roller, Sarah, and Megan Olsen,
Kelley Drye & Warren LLP. "Food Safety Modernization Act: Today and
Tomorrow." *Nutritional Outlook*, January/February 2012: 42–44. http://www
.kelleydrye.com/publications/articles/1547/_res/id=Files/index=0/.

171 **As for the good manufacturing practices:** Food and Drug Administration.
"Dietary Supplement Good Manufacturing Practices (CGMPs) and
Interim Final Rule (IFR) Facts," June 22, 2007 (archived). http://www
.fda.gov/Food/GuidanceRegulation/CGMP/ucm110858.htm. There
was a three-stage phase in process beginning in June 2008 for large
companies and finishing in June 2010 for companies with fewer than ten
employees.

172 **FDA's former director of Dietary Supplement Programs:** Phone conversation
with Daniel Fabricant, February 17, 2012.

172 **"that have more than five hundred employees":** Angela Pope, Office of
Dietary Supplements research practicum, June, 2014.

172 **about 600 . . . domestic dietary supplement manufacturers:** These figures
come from a July 19, 2014, interview with Charlotte Christin, acting director
of the FDA's Division of Dietary Supplement Programs. In addition, it is
likely the FDA is probably not always receiving notification when new dietary
ingredients hit the market. According to an article in the *New England Journal
of Medicine*, the FDA received notification of only 170 total new dietary
ingredients between 1994 and 2012—a far lower number than are suspected
to have been introduced. Cohen, Pieter. "Assessing Supplement Safety—The
FDA's Controversial Proposal." *New England Journal of Medicine* 366 (2012):
389–391. http://www.nejm.org/doi/full/10.1056/NEJMp1113325.

173 **The entire budget for the FDA's Division of Dietary Supplements:** Hurley, *Natural Causes*: 264.

173 **total of twenty-four full-time employees:** "Highlights from the 2013 Supplement Business Report." *Nutrition Business Journal,* December 9, 2013. http://newhope360.com/supplements/infographic-highlights-2013-supplement-business-report. To put this in context, in 2004 America's drug sales were 12 times greater than sales of supplements—and yet the FDA had almost 43 times as much money for drug regulation as it did for supplements, and almost 48 times as many people. "Dangerous Supplements Still at Large," *Consumer Reports.*

173 **proposed Dietary Supplement and Awareness Act:** "Dietary Supplement Access and Awareness Act," H.R. 3156, 109th Cong., 1st Sess. (2005). http://thomas.loc.gov/cgi-bin/query/z?c109:H.R.3156.IH:.

173 **2009 report from the US Government Accountability Office:** US Government Accountability Office. *Dietary Supplements: FDA Should Take Further Actions to Improve Oversight and Consumer Understanding*: 1.

173 **I asked Steve Mister:** These quotes from Steve Mister come from a phone conversation on May 7, 2013.

174 **"adulteration floods the market":** For example, when hurricanes battered Florida in 2005, much of the saw palmetto crop was wiped out. "All of a sudden, all this cheap supply starts coming from China," said Neal-Kababick. "Two things are wrong with that situation: first, there's a global shortage and now there are companies selling it cheaper. And second, saw palmettos don't grow in China. They're restricted to the southeastern United States. It'd be like growing a cactus at the North Pole." Interview with analytical chemist James Neal-Kababick, January 2013; "Bogus Ingredients Put Consumers' Health at Risk." NBCnews.com, June 14, 2007. http://www.nbcnews.com/id/19230748/#.Um_Gi3DBPD4.

175 **"What they didn't understand, though":** Hurley, *Natural Causes*: 88–89.

175 **"Much is at stake":** *Regulation of Dietary Supplements:* 65 (Statements of David Kessler, Commissioner, FDA).

176 **when I looked it up on ConsumerLab.com:** "Sea Buckthorn." ConsumerLab.com. http://www.consumerlab.com/tnp.asp?chunkiid=214734&docid=/epnat/herb_supp/Sea%20Buckthorn#ref29.

CHAPTER 9: Foods with Benefits

179 **most herbs' possible side effects:** Ginkgo biloba, for example, was found in an NCCAM-funded study to be "ineffective in lowering the overall incidence of dementia and Alzheimer's disease in the elderly [and] in slowing cognitive decline, lowering blood pressure, or reducing the incidence of hypertension." Its potential side effects, on the other hand, include

Notes

297

headache, nausea, gastrointestinal upset, diarrhoea, dizziness, allergic skin reactions, abnormal bleeding, and interaction with anticoagulant drugs. Oh, and two years of taking ginkgo extract appears to have caused tumours in rats. http://nccam.nih.gov/health/ginkgo/ataglance.htm.

179 "[T]erms like *conventional* and *alternative*": Offit, *Do You Believe in Magic?*: 98.

181 Nurses' Health Study: *The Nurses' Health Study Findings: Some Highlights.* http://www.channing.harvard.edu/nhs/?page_id=197

184 Launched in the early 1980s, MREs: Army Regulation 40-25 BUMEDINST 10110.6 AFI 44-141. "Medical Services Nutrition Standards and Education," June 15, 2001. http://www.apd.army.mil/pdffiles/r40_25.pdf.

185 "the best fed in the world": "Operational Rations of the Department of Defense." Natick Pam 30–25, 8th ed., April 2010: 3.

185 "if their preferences were taken into consideration": Ibid.: 7.

187 many fortified foods and vitamins contain "overages": Borenstein, Benjamin, et al. "Fortification and Preservation of Cereals." In *Breakfast Cereals and How They Are Made*, eds. Robert B. Fast and Elwood F. Caldwell. St. Paul, MN: American Association of Cereal Chemists, 1990: 279.

188 clip from the TV show *Rock Center*: http://www.nbcnews.com/id/21134540/vp/45298505#45298505.

191 chemist Linus Pauling erroneously claimed: Barrett, Stephen. "The Dark Side of Linus Pauling's Legacy." QuackWatch.com, October 23, 2008. http://www.quackwatch.com/01QuackeryRelatedTopics/pauling.html.

192 some sort of beneficial effect after exercise: Reynolds, Gretchen. "Why Vitamins May Be Bad for Your Workout." *New York Times*, Well blog, February 12, 2014. http://well.blogs.nytimes.com/2014/02/12/why-vitamins-may-be-bad-for-your-workout.

192 "most persuasive evidence . . . diet-cancer epidemiological literature": Hercberg, Serge. "The History of β-Carotene and Cancers: From Observational to Intervention Studies. What Lessons Can Be Drawn for Future Research on Polyphenols?" (supplement) *American Journal of Clinical Nutrition* 81 (2005): 219S.

192 when Peto's paper: Omenn, Gilbert S. "Chemoprevention of Lung Cancer: The Rise and Demise of Beta-Carotene." *Annual Review of Public Health* 19 (1998): 74.

192 Data from the first trial: Blot, W. J., J. Y. Li, P. R. Taylor, W. Guo, S. Dawsey, G. Q. Wang, C. S. Yang, S. F. Zheng, M. Gail, J. Y. Li, et al. "Nutrition Intervention Trials in Linxian, China: Supplementation with Specific Vitamin/Mineral Combinations, Cancer Incidence, and Disease-Specific Mortality in the General Population." *Journal of the National Cancer Institute*

85, no.18 (September 15, 1993): 1483–1492. http://www.ncbi.nlm
.nih.gov/pubmed/8360931.

193 **9 percent lower incidence of death:** This particular treatment arm of the
Linxian trial gave 15 mg/day of beta-carotene, 30 mg/day of vitamin E, and
50 mcg/day of selenium. Ibid.

193 **"persistently low intake of several micronutrients":** Ibid.

193 **Alpha-Tocopherol and Beta-Carotene (ATBC) Cancer Prevention Trial:**
National Cancer Institute. "Alpha-Tocopherol, Beta-Carotene Cancer
Prevention (ATBC) Trial," posted 7/22/2003. http://www.cancer.gov/news
center/qa/2003/atbcfollowupqa; original study: http://www.nejm.org/doi/full/
10.1056/NEJM199404143301501. ATBC trial design: The doses in the ATBC
study were 50 mg/day of alpha-tocopherol and 20 mg/day of beta-carotene
(20 mg/day of beta-carotene in supplement form is the equivalent of about
33,340 IU of vitamin A as retinol—about ten times the current RDA of
3,000 IU of preformed vitamin A). 20 mg/day bc = 10 mcg retinol. Basis for
conversion math: 1 mg beta-carotene = 1667 IU vitamin A, so 20 mg
beta-carotene = 33,340 IU. http://www.utsandiego.com/uniontrib/20060228/
news_lz1c28qanda.html. RDA: http://ods.od.nih.gov/factsheets/VitaminA
-HealthProfessional/.

193 **16 percent *increase* in . . . lung cancer:** National Cancer Institute, "Alpha-
Tocopherol, Beta-Carotene Cancer Prevention (ATBC) Trial." ATBC trial
results: There were 474 cases of lung cancer in the beta-carotene group
compared with 402 in the control, which led to 302 deaths compared with
262. Men who took both beta-carotene *and* vitamin E had similar results,
but men who just took the vitamin E had 32 percent fewer cases of prostate
cancer and 41 percent fewer deaths from prostate cancer. But before you
conclude vitamin E is the good guy, consider this: death from hemorrhagic
stroke increased 50 percent in the men taking the vitamin E supplements,
particularly if they had high blood pressure. In fact, vitamin E is consistently
confusing: the SELECT Prostate Cancer Prevention Trial was the largest-
ever prostate cancer prevention trial, inspired by nonclinical research
suggesting that selenium and vitamin E might decrease the risk
of developing prostate cancer by 60 and 30 percent, respectively. In 2011,
after an average of seven years of treatment (the men took the vitamins for
five and a half years and then the researchers waited another year and a
half to measure any lingering effects), the men taking only vitamin E
had 17 percent *more* cases of prostate cancer than the men who had been
taking placebos (76 cases per thousand men versus 65 cases per thousand
men). "Study Shows Increased Prostate Cancer Risk from Vitamin E
Supplements." *Science Daily*, October 12, 2011. http://www.sciencedaily
.com/releases/2011/10/111011163049.htm; SELECT trial: National
Cancer Institute. "SELECT Prostate Cancer Prevention Trial,"
updated 10/12/2011. http://www.cancer.gov/clinicaltrials/noteworthy-trials/
select/page1.

193 **CARET study:** CARET's intervention group received 30 mg/day of beta-carotene and 25,000 IU/day of preformed vitamin A (in the form of retinyl palmitate).

194 **28 percent *increase* in lung cancer:** National Cancer Institute. "Antioxidants and Cancer Prevention: Fact Sheet," last reviewed 7/28/2004. http://www .cancer.gov/cancertopics/factsheet/prevention/antioxidants. Original paper: Omenn, Gilbert S., et al. "Effects of a Combination of Beta-Carotene and Vitamin A on Lung Cancer and Cardiovascular Disease." *New England Journal of Medicine* 334, no. 18 (May 2, 1996): 1150–1155.

194 **35–40 percent more likely to develop lung cancer:** Berg, Barbara. "Enduring Effects of Lung-Cancer Supplement Study." Fred Hutchinson Cancer Research Center, December 16, 2004. http://www.fhcrc.org/en/news/ center-news/2004/12/dietary-supplements.html.

194 **"It was as if Dr. Jekyll became Mr. Hyde":** Hercberg, "The History of β-Carotene and Cancers": 221S.

194 **Physicians' Health Study I:** Hennckens, C. H., et al. "Lack of Effect of Long-Term Supplementation with Beta-Carotene on the Incidence of Malignant Neoplasms and Cardiovascular Disease." *New England Journal of Medicine* 334, no. 18 (May 2, 1996): 1145–1149. http://www.ncbi.nlm.nih.gov/ pubmed/8602179?dopt=Abstract.

194 **beta-carotene and the Women's Health Study:** Lee, I. M., et al. "Beta-Carotene Supplementation and Incidence of Cancer and Cardiovascular Disease: The Women's Health Study." *Journal of the National Cancer Institute* 91, no. 24 (December 15, 1999): 2102–2106. http://jnci.oxfordjournals.org/ content/91/24/2102.full.pdf+html.

194 **Treatment details: PHS I and Women's Health Study:** PHS I's intervention group took 50 mg of beta-carotene every other day for twelve years; the Women's Health Study gave 50 mg of beta-carotene every other day for two years and followed subjects for an additional two years.

194 **beta-carotene in supplements and beta-carotene . . . in food:** "The New Recommendations for Dietary Antioxidants: A Response and Position Statement by the Linus Pauling Institute," May 2000. http://lpi.oregonstate .edu/s-s00/recommend.html.

195 **SU.VI.MAX study:** Hercberg, Serge, et al. "The SU.VI.MAX Study: A Randomized, Placebo-Controlled Trial of the Health Effects of Antioxidant Vitamins and Minerals." *Annals of Internal Medicine* 164 (2004): 2335–2342. http://www.unpieddansleplat.fr/menu_gauche/suvimax.pdf. But not women! The researchers hypothesized that this might have been because the women had higher levels of antioxidants, particularly beta-carotene, at the beginning of the study than the men. The SU.VI.MAX study gave 6 mg/day of beta-carotene compared with 20 mg/day for the ATBC trial, 30 mg/day for the CARET trial, and 50 mg every other day for PHS I.

195 **results of the Physicians' Health Study II:** The PHS II found no associations
    between multivitamins and cardiovascular disease (or cognitive function)—
    but that's a controversial issue as well. Indeed, according to a 2003 report
    from the US Preventive Services Task Force, "Four cohort studies
    analyzed the relationship between the use of multivitamins and
    cardiovascular disease. One good quality study reported a significant
    reduction in coronary events with multivitamin use. 2 good quality studies
    reported no significant effect on mortality, and a fair quality trial reported an
    increase in all-cause mortality among men."

196 **more than half of Americans report taking dietary supplements:** Bailey,
    Regan, Victor Fulgoni, Debra Keast, and Johanna Dwyer. "Examination of
    Vitamin Intakes Among US Adults By Dietary Supplement Use." *Journal of
    the Academy of Nutrition and Dietetics*, vol. 112, no. 5 (May 2012): 657–663.

196 **this phenomenon is referred to as "hidden hunger":** http://globalfoodforthought
    .typepad.com/global-food-for-thought/2012/12/interview-with-dr-klaus-
    kraemer-on-tackling-malnutrition-and-micronutrient-deficiencies.html.

196 **the American Heart Association, . . . do not recommend:** US Preventive Task
    Force. "Routine Vitamin Supplementation to Prevent Cancer and
    Cardiovascular Disease, Recommendations and Rationale," June 2003.
    http://www.aafp.org/afp/2003/1215/p2422.html.

196 **folic acid . . . vitamin B12:** Offit, *Do You Believe in Magic?*: 103; Hurley,
    Natural Causes: 260.

197 **"The message is simple":** Guallar, Eliseo, et al. "Enough Is Enough: Stop
    Wasting Money on Vitamin and Mineral Supplements." *Annals of Internal
    Medicine* 159, no. 12 (December 17, 2013): 850–851. http://annals.org/article.
    aspx?articleid=1789253.

CHAPTER 10: The Nutritional Frontier

199 **$4.7 billion in annual sales:** http://globalnews.amway.com/pressroom/
    news-release/news/amway-breaks-ground-on-38-million-nutrilite-botanica
    l-concentrate-manufacturing-facility-in-quincy-washington.

199 **vitamins do similar things for plants:** Plant details based on conversations
    with Gene Lester and Janet Slovin at USDA.

200 *some* **of the phytochemicals under study:** American Cancer Society fact page
    on phytochemicals: http://www.cancer.org/treatment/treatmentsandside
    effects/complementaryandalternativemedicine/herbsvitaminsandminerals/
    phytochemicals.

201 **an infusion of rusty nails:** This idea is actually not that wacky. Cooking food
    in iron pots substantially increases the meal's iron content. That's the
    inspiration behind the Lucky Iron Fish project: a nutritional intervention
    effort, currently focused in Cambodia, that tries to combat widespread iron

deficiencies in the developing world by giving families a "good luck" iron fish that they are instructed to drop into the pot while sterilizing water or preparing meals. The organization first tried providing blocks of iron to cook with, but switched to a good-luck symbol when it noticed that the local women tended to use the blocks as doorstops. http://www.theatlantic .com/magazine/archive/2014/01/an-iron-fish-in-every-pot/355742/.

201 **Nutrilite top sellers, its Double X:** Amway online catalog: http://www. amway.com/Shop/Product/Product.aspx/NUTRILITE-DOUBLE-X-Vitamin-Mineral-Phytonutrient-31-Day-Supply-with-Case?itemno=A4300.

202 **study published in 2000 in *Nature*:** Eberhardt, Marian, et al. "Nutrition: Antioxidant Activity of Fresh Apples." *Nature* 405 (June 22, 2000): 903–904.

202 **"dirty chromatogram":** Visit to Nutrilite (Lakeview, CA), July 5, 2012.

203 **My tour guide for the day:** I met with Saito during my visit to Nutrilite's headquarters on July 5, 2012.

205 **Or consider berberine:** National Center for Complementary and Alternative Medicine. "Herbs at a Glance: Goldenseal." http://nccam.nih.gov/health/ goldenseal.

206 **2011 study on broccoli:** Clarke, J. D., et al. "Bioavailability and Inter-Conversion of Sulforaphane and Erucin in Human Subjects Consuming Broccoli Sprouts or Broccoli Supplement in a Cross-Over Study Design." *Pharmacological Research Journal* 64, no. 5 (2011): 456–463.

206 **different from those found in food:** Williamson, Gary, and Claudine Manach. "Bioavailability and Bioefficacy of Polyphenols in Humans. II. Review of 93 Intervention Studies," (supplement) *American Journal of Clinical Nutrition* 81 (2005): 243S–255S.

206 **freeze-dried whole tomato powder:** Campbell, Jessica K., et al. "Tomato Phytochemicals and Prostate Cancer Risk," Supplement 12. *Journal of Nutrition* 134 (2004): 3486S–3492S.

207 **conference on food, nutrition, and cancer:** Ibid.: 3486S. http://jn.nutrition. org/content/134/12/3486S.long.

207 **not the types of issues that lend themselves:** Jacobs, David, and Lyn Steffen. "Nutrients, Foods and Dietary Patterns as Exposures in Research: A Framework for Food Synergy," (supplement) *American Journal of Clinical Nutrition* 78 (2003): 508S.

208 **5 percent of American adults under age fifty:** *Report of the Dietary Guidelines Advisory Committee on the Dietary Guidelines for Americans, 2010*: 133–134.

209 **the people who use dietary supplements:** Bailey, Fulgoni, Keast, and Dwyer, "Examination of Vitamin Intakes Among US Adults by Dietary Supplement Use": 657.

209 data from . . . NHANES: *Report of the Dietary Guidelines Advisory Committee on the Dietary Guidelines for Americans, 2010*: 137.

209 Blood and urine tests . . . Centers for Disease Control: Centers for Disease Control and Prevention. *National Report on Biochemical Indicators of Diet and Nutrition in the U.S. Population 2012.* http://www.cdc.gov/nutritionreport. The CDC's report, which was based on NHANES data from 1999 to 2006, found that 10.5 percent of Americans were deficient in vitamin B6, 7 percent of American men were deficient in vitamin C, and 4 percent of older adults were deficient in B12. As for vitamin D, while only 3 percent of non-Hispanic white people were deficient, 12 percent of Mexican Americans and 31 percent of non-Hispanic black people were deficient (in which deficiency was defined as having a serum 25-hydroxyvitamin D level of less than 30 nmol/L).

212 Most scientific organizations are extremely dubious: National Institutes of Health. "Genetic Testing." *Genetics Home Reference*, June 9, 2014. http://ghr.nlm.nih.gov/handbook/testing%3Fshow=all#validtest.

212 genetic variations called single nucleotide polymorphisms: National Institutes of Health. "What Are Single Nucleotide Polymorphisms?" *Genetics Home Reference*, June 9, 2014. http://ghr.nlm.nih.gov/handbook/genomicresearch/snp.

212 every person has about 10 million SNPs: Winstead, Edward. "GNN's Genome Glossary." Genome News Network. http://www.genomenewsnetwork.org/resources/glossary/; National Institutes of Health, "What Are Single Nucleotide Polymorphisms?"

213 every baby has about fifty to seventy new SNPs: Conversation with José Ordovas, senior scientist and director, Nutrition and Genomics Laboratory, chair, Functional Genomics Core Scientific Advisory Committee, Functional Genomics Core, Tufts University Jean Mayer USDA Human Nutrition Research Center on Aging, March 25, 2013.

213 in a way that can cause disease: National Institutes of Health, "What Are Single Nucleotide Polymorphisms?"

213 "the question should not be, 'Is this a genetic disorder?'": American Dietetic Association and the National Genetics Education and Development Centre. "Health and Disease: The Result of Interactions between Genes and Environment." http://www.nchpeg.org/nutrition/index.php?option=com_content&view=article&id=412&Itemid=559.

213 caffeine increases your risk of bone loss: Harris, S. S., and B. Dawson-Hughes. "Caffeine and Bone Loss in Healthy Postmenopausal Women." *American Journal of Clinical Nutrition* 60, no. 4 (1994): 573–578. http://ajcn.nutrition.org/content/60/4/573.abstract.

214 "The time is quickly approaching": Combs, *The Vitamins*: 486.

214 **"It's like someone rings your doorbell":** Conversation with José Ordovas, March 25, 2013.

215 **The official daily rations per person:** Roseboom, Tessa, Jan van der Meulen, Anita Ravelli, Clive Osmond, David Barker, and Otto Bleker. "Effects of Prenatal Exposure to the Dutch Famine on Adult Disease in Later Life: An Overview." *Molecular and Cellular Endocrinology* 185 (2001): 94.

215 **famine reached its peak in April 1945:** Hoek, H. W., A. S. Brown, and E. Susser. "The Dutch Famine and Schizophrenia Spectrum Disorders." *Society of Psychiatry and Psychiatric Epidemiology* 33 (1998): 374.

215 **Dutch Hunger Winter killed at least 22,000 people:** Trienekens, Gerard. "The Food Supply in The Netherlands during the Second World War." In *Food, Science, Policy and Regulation in the Twentieth Century: International and Comparative Perspectives*, eds. David F. Smith and Jim Phillips. London: Routledge, 2000: 127.

216 **children of the malnourished women:** The effects were not all negative. One of the earlier Dutch Hunger Winter studies, published in 1976, found that young men whose mothers had been exposed to the famine late in their pregnancies actually weighed less as adults than young men whose mothers had experienced the famine earlier in their pregnancies. Ravelli, Gian-Paolo, Zena Stein, and Mervyn Susser. "Obesity in Young Men after Famine Exposure In Utero and Early Infancy." *New England Journal of Medicine* 295 (August 1976): 349–353.

216 **effects . . . seen in Gambian children:** Waterland, Robert, et al. "Season of Conception in Rural Gambia Affects DNA Methylation at Putative Human Metastable Epialleles." *Plos Genetics*, December 23, 2010. http://www .plosgenetics.org/article/info%3Adoi%2F10.1371%2Fjournal.pgen .1001252.

216 **twofold risk of developing schizophrenia:** Susser, Ezra, Richard Neugebauer, Hans W. Hoek, Alan S. Brown, Shang Lin, Daniel Labovitz, and Jack M. Gorman. "Schizophrenia After Prenatal Famine: Further Evidence." *Archives of General Psychiatry* 53 (1996): 25–31; Hoek, Brown, and Susser, "The Dutch Famine and Schizophrenia Spectrum Disorders": 377.

217 **grandchildren of smokers . . . higher risk of asthma:** Painter, R., C. Osmond, P. Gluckman, M. Hanson, D. Phillips, and T. Roseboom. "Transgenerational Effects of Prenatal Exposure to the Dutch Famine on Neonatal Adiposity and Health in Later Life." *BJOG: An International Journal of Obstetrics and Gynaecology*, 115 (2008): 1243; Zeisel, Steven H., et al. "Epigenetic Mechanisms for Nutrition Determinants of Later Health Outcomes," (supplement) *American Journal of Clinical Nutrition* 89 (2009): 1488S–1493S. http://www.ncbi.nlm.nih.gov/pmc/articles/PMC2677001/.

217 **"We are only beginning to appreciate":** Roseboom, T. J., and E. D. Watson. "The Next Generation of Disease Risk: Are the Effects of Prenatal

Nutrition Transmitted Across Generations? Evidence from Animal and
Nutrition Studies." *Placenta* 33 (2012): e40–e44.

218 **thought to have connections to diet:** DeBusk, Ruth, et al. "Nutritional
Genomics in Practice: Where Do We Begin?" *Journal of the American Dietetic
Association* 105, no. 4 (April 2005): 591.

218 **disease risk can be passed down to grandchildren:** Daxinger, Lucia, and
Emma Whitelaw. "Understanding Transgenerational Epigenetic Inheritance
via the Gametes in Mammals." *Nature Reviews: Genetics*, March 2012:
153–162; Zeisel et al. "Epigenetic Mechanisms for Nutrition Determinants
of Later Health Outcomes": 1488S–1493S.

218 **the 1996 adaptation of *Romeo and Juliet*:** Carey, Nessa. *The Epigenetics
Revolution: How Modern Biology Is Rewriting Our Understanding of Genetics,
Disease and Inheritance.* New York: Columbia University Press, 2012: 7.

220 **"profoundly change our understanding of inheritance":** Daxinger and
Whitelaw. "Understanding Transgenerational Epigenetic Inheritance via the
Gametes in Mammals": 156.

221 **a highly publicized 2003 paper . . . Duke University:** Waterland, Robert A.,
and Randy L. Jirtle. "Transposable Elements: Targets for Early Nutritional
Effects on Epigenetic Gene Regulation." *Molecular and Cellular Biology* 23,
no. 15 (2003): 5293–5300.

221 **a photograph of two mice:** Kollias, Helen. "Research Review: Epigenetics."
PrecisionNutrition.com. http://www.precisionnutrition.com/
epigenetics-feast-famine-and-fatness.

222 **betaine, a chemical:** WebMD. "Find a Supplement: Betaine Anhydrous."

223 **our understanding of how diet affects an embryo:** If you want further proof
that it is important, however, scientist Nessa Carey suggests looking no
further than the honeybee. Queen bees start off no different from the
thousands of worker bees in their hives, and yet once matured they can be
up to twice as large—and typically live up to 20 times longer than the other
bees. (They also have no stingers or pollen baskets and are the only female
bees that can reproduce.) So what causes the differences? Royal jelly, a
complex mixture of fats, proteins, vitamins, and amino acids that a
particular class of worker bees can secrete from glands in their heads. Most
larvae only receive royal jelly for about three days; they turn into normal
worker bees. Several special larvae, on the other hand, continue to be fed
royal jelly. Those larvae grow up to be the queens. Carey, *The Epigenetics
Revolution*: 283–4.

223 **effects on our microbiomes:** There is increasing evidence today that what we
eat may actually affect our moods (if not long-term personality traits),
perhaps by affecting the production of mood-affecting neurotransmitters
like serotonin, or by influencing the population of microbes in our guts,
which themselves may have effects on neural development and brain

chemistry that are not yet understood. Carpenter, Siri. "That Gut Feeling." *American Psychological Association* 43, no. 8 (September 2012): 50. http://www.apa.org/monitor/2012/09/gut-feeling.aspx.

223 **"arguably the most intimate connection that humans have":** Pray, Leslie, Laura Pillsbury, and Emily Tomayko, rapporteurs. *The Human Microbiome, Diet and Health: Workshop Summary.* Washington, DC: National Academies Press, 2012: 1.

223 **its population is nearly unbelievably large:** Differences in gut flora might actually account for some differences between human cultures. For example, a 2010 study done by researchers from France's Centre National de la Recherche Scientifique suggested that many Japanese people are more efficient than other people at digesting carbohydrates from seaweed because of particular populations in their gut microflora. Eisenstein, Michael. "Of Beans and Genes." *Nature* 468 (2010): S15.

223 **between 1 and 2 percent of our body weight:** Baylor College of Medicine Department of Molecular Virology and Microbiology. "The Human Microbiome Project." https://www.bcm.edu/departments/molecular-virology-and-microbiology/microbiome.

223 **microbiome is estimated to have about 3.3 *million*:** Pray, Pillsbury, and Tomayko, *The Human Microbiome, Diet and Health: Workshop Summary*: 44.

EPILOGUE

227 **"We are complex organisms":** Carey, *The Epigenetics Revolution*: 113.

229 **where cooking can increase:** *European Journal of Clinical Nutrition* 56 (2002): 425–430.

APPENDIX A: The Vitamins

235 **The Vitamins:** Most of the background information about the vitamins comes from either Combs, Gerald Jr., *The Vitamins: Fundamental Aspects in Nutrition and Health, 4th Edition,* or the Medline information sheets from the National Library of Medicine/National Institutes of Health.

236 **"lifting from his ear a perfect skin cast":** Feeney, Robert E. *Polar Journeys: The Role of Food and Nutrition in Early Exploration.* Washington, DC, and Fairbanks: American Chemical Society/University of Alaska Press, 1998: 240, 169.

236 **It's most often seen in severe alcoholics:** "Beriberi." MedlinePlus: http://www.nlm.nih.gov/medlineplus/ency/article/000339.htm.

237 **Formerly known as vitamin G:** Baker, Mary Zoe. "Riboflavin Deficiency." Medscape: http://emedicine.medscape.com/article/125193-overview.

238 It's also in coffee beans: Ottaway, P. Berry, ed. *The Technology of Vitamins in Food.* Glasgow: Blackie Academic & Professional, 1993: 31.

238 ovaries of coldwater fish: Combs, *The Vitamins*: 340.

238 not present in common heavily refined foods: Ottaway, ed., *The Technology of Vitamins in Food*: 35.

239 moldy cheeses or cheese rinds: Ibid.: 34.

240 the rate of neural tube defects in the United States: "The Ups and Downs of Folic Acid Fortification." *HealthBeat*, Harvard Medical School, 2008. http://www.health.harvard.edu/newsweek/the-ups-and-downs-of-folic-acid-fortification.htm.

242 Symptoms of a B12 deficiency: Frankenburg, *Vitamin Discoveries and Disasters*: 69.

243 It's also abundant in sauerkraut: Ibid., p. 80.

243 it also has industrial uses: This is a directly quoted list: Counsell, J. N. "Vitamins as Food Additives." In *The Technology of Vitamins in Food*, ed. P. Berry Ottaway: 144.

245 Vitamin D is also used as a rat poison: "Rodenticide Topic Fact Sheet." National Pesticide Information Center. http://npic.orst.edu/factsheets/rodenticides.html.

245 Vitamin E: A lot of this is from Shi, John, G. Mazza, and Marc Le Maguer, eds. *Functional Foods: Biochemical and Processing Aspects, Vol. 2.* Boca Raton: CRC Press, 2002: 8.

INDEX